Finding the Voice of the River

Gary J. Brierley

Finding the Voice of the River

Beyond Restoration and Management

Gary J. Brierley
School of Environment
University of Auckland
Auckland, Aotearoa, New Zealand

ISBN 978-3-030-27067-4   ISBN 978-3-030-27068-1   (eBook)
https://doi.org/10.1007/978-3-030-27068-1

This Palgrave Pivot imprint is published by the registered company Springer Nature
Switzerland AG.
The registered company address is: Gewerbestrasse 11, 6330 Cham, Switzerland

To those who help rivers find their voice

PREFACE

With societal relations to rivers at its core, this book expresses a river-centric view of the world—a plea to live with rivers, and with each other, in compassionate and respectful ways.

Contemporary times are characterized by profound unease, unrest and disquiet, troubled by the extent and pace of change on the one hand, and the ineptitude of political institutions and governance arrangements that purportedly act in the interests of society on the other. Despite incredible potential, equitable practices that serve the greater good have been pushed aside.

Across much of the planet, rivers have been marginalized from the lives of much of human society—they have been 'othered'. Physical and psychological separation accentuates disconnection, unconsciously furthering neglect. Whether piped and built over in urban areas, channelized and canalized to suppress their voice, or treated with disdain as gutters, drains and sewers, day to day relations with many contemporary watercourses limit positive experiences to our imagination. More frequently, connections to riverscapes reflect idealized natures expressed within the realms of virtual reality, gaming worlds or theme parks.

Despite their fundamental importance to human society, rivers have been treated with such disdain that they are now in a perilous state. In these crazy times of denial and doubt that shape "The Great Derangement" (Ghosh 2016), efforts to improve river health start by acknowledging that human endeavours created these problems, recognizing that it is our responsibility to address them. Two very different approaches to river repair are available, framed here as Medean (competitive) and Gaian (cooperative) options. Rather than 'managing' rivers to achieve particular anthropogenic goals, this book outlines a more-than-human approach to 'living with living rivers', conceptualizing rivers as sentient entities that are allowed to express their own voice. Adoption of a duty of care conceptualizes the river as a living entity—an indivisible whole, from the mountains to the sea—minimizing harm and embracing measures to improve its health, respecting the rights of the river itself.

As environmental conditions deteriorate, populations grow and locked-in path dependencies become harder to revoke, choices for prospective futures become increasingly more limited, constraining prospects for future generations. The warning signs are clear. Rather than sleepwalking into oblivion, we know the consequences of our actions, despite significant frustration at the ineptitude of societal and institutional responses.

This book is written for a generalist audience. It is a Geography book, written by a Physical Geographer. Being disruptive and working across boundaries are liberating traits of geographic enquiry. Although the book has an environmental focus, it is framed as an argument in popular philosophy, contemplating the agency of rivers as sentient entities—place-beings. It relates situated place-based knowledges (concerns for a given river) to global considerations (concerns for all rivers).

Deliberations are assuredly upbeat and aspirational, applying a strategy of hope that envisages the glass as half-full rather than half-empty, reflecting that it pays to dream, engaging with the world as it could be.

The book has a linear structure, and is intended to be read from front to back rather than dropping in and out of chapters (though boxes present occasional digressions, as examples, personal anecdotes or river stories). Chapter 1 outlines what it means to 'Find the Voice of the River' (the sentient river). Chapter 2 presents an overview of societal relations to rivers, outlining perceptions and uses of rivers, and impacts upon them (the socio-ecological river). Chapter 3 juxtaposes a competitively framed (Medean) approach to river repair relative to a cooperatively framed approach that incorporates 'Finding the Voice of the River' (a Gaian lens) (from managing to living with rivers). Finally, Chap. 4 applies a river health metaphor and an analogy with medical practice to present a rallying call to look after rivers in more respectful ways through place-based approaches to river repair (finding and enacting the voice of the river).

How societies choose to live with rivers is a microcosm of their relations with the world and each other. These are big issues to consider in a short book.

Auckland, Aotearoa, New Zealand   Gary J. Brierley

REFERENCE

Ghosh, A. (2016). *The Great Derangement: Climate Change and the Unthinkable*. Chicago: University of Chicago Press.

Acknowledgements

This book feels like a life-long quest—something that was waiting to emerge, even if it didn't know it was going to do so. There were many instances in which it felt as though it may not happen, so it's an enormous relief to be writing this particular section!

I owe an enormous debt of gratitude to a host of people who have supported this product, often unknowingly. Countless conversations along rivers and sharing of river experiences contributed to this book. My family put up with a very particular take on the world, and I want to thank them first—my wife, ME; our kids, Zachary and Whittam; my mum, Joyce; and my siblings, Sara and Mark.

The first draft of this book now feels an eternity ago, but helpful and incisive comments by the reviewer, editor and various close friends and colleagues helped to transform that initial version to a much better product (I hope). Marc, Carola, Richie and Anne—your comments helped enormously. ME, Carola and Mark helped me clean up a bunch of things at the end of this journey. Ian and Pete helped me work out which direction to head in taking initial steps. I also thank Rachel and Jo at Palgrave Macmillan for their guidance and assistance with this book, and Hemalatha Arumugam for working through the proofs so efficiently.

In keeping with the theme of the book, I want to reflect upon how various rivers have influenced my life. How has their voice shaped this book?

Unlike many other people, I don't feel that I'm 'of' a particular river. Rather, several rivers have played key roles in my life, and they are surely a part of me. Here's a snapshot of my story, as told through connections to rivers (cf., Duncan 2001).

The family farm was located at the confluence of the Roch and the Irwell, near Bury in northwest England. Given the industrial heritage of the region, my birth river was smelly and slimy, its voice a scream of anguish. Decades later, at the first River Festival in Brisbane, Australia, in 1999, the Mersey Basin Campaign was the first winner of the Thiess International Riverprize. Remarkable improvements in water quality had greatly increased fish populations across much of the catchment. It was weird to see aerial images of the former family farm site during the awards ceremony.

My first rivers-related goosebump moments occurred along semi-arid watercourses in Tunisia and the West Bank during my undergraduate days. Their voice oozed history and the fundamental importance of water. I'll forever appreciate the open-mindedness of my lecturers at Durham University who allowed a group of us to rent a car and explore rivers around Kairouan (this would never happen today!). A few months later, a watery encounter in a Berber camp near Jericho resulted in a variant of paratyphoid—I can still see those wiggling bugs in the bottom of the rusty can.

Glaciated riverscapes of the Squamish Valley in British Columbia, western Canada, transformed my life. Postgraduate experiences include reflections upon mobile boulders, dramatic rapids and enormous log jams along a river that needed no encouragement to loudly proclaim its voice.

Once the ancient, orange-red landscapes of Australia got into my blood, in ways that felt as though they had always been there, there was no going back. These rivers were crying out for management practices that respected their inherent diversity. Their voices were neglected and unheard, at least by the authorities charged to look after them.

Walking down tailings deposits at a mine-site in Papua New Guinea in the mid-1990s, every splash from the toxic stream made a hole in my clothing. Such disrespect for the river provided a frame of reference like no other—the voice of the river was utterly disregarded.

For the last 15 years, I have lived and worked in the geomorphological paradise of Aotearoa New Zealand. Inspired by professional collaborations and insights into Māori relationships to landscapes, I now see rivers in a different light. Assertions of river rights offer genuine prospect for more-than-human deliberations that encourage rivers to express their voice.

Reference

Duncan, D. J. (2001). *My Story as Told by Water*. San Francisco: Sierra Book Clubs.

CONTENTS

LIST OF TABLES

LIST OF BOXES

CHAPTER 1

What Does It Mean to Find the Voice of the River?

Abstract Rivers have played a fundamental role in the development of human society. Despite their importance to societal wellbeing, rivers continue to be treated with disdain. Many rivers are in a perilous state. An era of river repair has begun, but it is not working in the manner or scale that is required. Drawing on Māori perspectives from Aotearoa New Zealand and other global visions of waterways, *Finding the Voice of the River* builds upon emerging approaches to river rights that express a more-than-human lens. Rather than managing a river to a particular norm, a culture and duty of care emphasizes concerns for reciprocity, coevolution and mutual interdependence, living with rivers as living entities. Such framings envisage harmonious relations to rivers, and each other.

Keywords River condition • River management • Socio-natures • Environmental values • Traditional knowledges • More-than-human • River rights

1.1 A River-Centric View of a Healthier World

The river has taught me to listen; you will learn from it, too. The river knows everything; one can learn everything from it. Herman Hesse (1951, *Siddhartha*, p. 105)

© The Author(s) 2020
G. J. Brierley, *Finding the Voice of the River*,
https://doi.org/10.1007/978-3-030-27068-1_1

1

Contemporary society is in the midst of an existential crisis. On the one hand, the Information Age presents profound egalitarian potential associated with the rapid pace of technological and social change. On the other, there is profound disquiet and increasing alarm as existing governance arrangements fail to come to terms with environmental, socio-political and cultural problems at local, regional, national and global scales. Although remarkable insight and evidence have been achieved, governance arrangements are unable to address concerns for climate change, population growth, increasing consumption, resource depletion and the perils faced as planetary boundaries and tipping points are breached (e.g. Rockström et al. 2009; Steffen et al. 2015). Systemic crises have intellectual, moral and spiritual dimensions (Capra 1982). They are not issues of managerial inefficiency that can be addressed one at a time. Rather, they reflect a profound imbalance in thoughts and feelings, values and attitudes, and social and political structures. Different ways of thinking and living are required to address these issues of conscience and consciousness, recognizing that human survival and wellbeing are innately enmeshed within the fate of the environment. The earlier warnings of 'Limits to Growth' and 'Only One Earth' ring loud and clear (Friends of the Earth 1972; Meadows et al. 1972).

Just as the health of a canary in a cage was used to provide guidance into prospective disasters in underground mines, this book considers river health as a measure of the ways human society is living with our planet. As impacts of human disturbance accumulate and are accentuated on valley floors, rivers provide a powerful focal point in considering ways to address prevailing environmental crises. Rivers play fundamental roles in our interconnected and interdependent lives. We depend on rivers, as rivers depend on us. They are not only the lifeblood of the land; they are also the lifeblood of society. Although rivers are venerated across the planet, they have been desecrated in just about every conceivable way, imperilling the health and viability of aquatic ecosystems (Millennium Ecosystem Assessment 2005). Sadly, in many instances, the canary in the cage is decidedly unwell. Rather than singing sweet melodies, the voice of the river is screaming in angst. Alarm bells are ringing.

An alternative way of thinking is required, repositioning human endeavour and societal relations to the world. As a starting point, it must be recognized that:

1. We are 'of' the Earth—the Earth made us, and we are fundamentally dependent upon it (Dartnell 2018).
2. Social and environmental crises are not someone else's problem and no-one else is going to look after this place for us. It's up to us.
3. It is time to cast aside denial, despair and excuses, placing environmental concerns at the front and centre of political agendas.

The struggle to overthrow our life-denying system has begun, as a growing voice of activist movements seeks to defend our life-support systems (Monbiot 2019). It took several thousand years to develop the technical mastery to tame and control rivers. In recent decades, moves towards an era of river repair have brought about some remarkable transformations in river condition, exemplified by marked improvements in water quality, environmental flow allocations and countless restoration initiatives. The process of river repair has taken different forms in different parts of the world, turning the tide of environmental degradation in some instances as contemporary societies live with rivers in fundamentally different ways to past eras (Brierley and Fryirs 2008). Indeed, many past practices would be considered unconscionable, almost inconceivable today. Much depends upon contextual considerations at a given place, as historical factors and the trajectory of river adjustment constrain what is realistically achievable for each river system (see Dufour and Piégay 2009).

However, this book contends that the ethos and mentality that underpins steps taken towards an era of river repair will not achieve the scale and rate of transformation that is required. Rather than accentuating or limiting damage to the river, a different ethos and approach puts the interests of the river front and centre, instilling efforts to allow each river to express its own voice. Increasing commitment to the Rights of Nature and a growing Earth Jurisprudence movement presents a compelling backbone for such prospects (Bosselmann 2008; Boyd 2017; Chapron et al. 2019; O'Donnell and Talbot-Jones 2018). For example, the Whanganui River in Aotearoa New Zealand is now a legal entity—it has its own rights (e.g. Morris and Ruru 2010; Ruru 2018).

With rivers at its heart, this book outlines what a more-than-human approach to living with rivers as living entities looks like, and prospects to achieve it. *Finding the Voice of the River* views rivers as living and sentient entities—place-beings with moral standing. Such relations are simultaneously material and spiritual. They are ecocentric, not anthropocentric. An ecosystem approach views humans as part of nature, working with the river, not managing it to a particular norm. Rather than focussing upon short-term projects for a few flagship rivers, typically tied to budget and political cycles, concerns for social and environmental justice emphasize collective engagement in efforts to protect and enhance core values (things that matter) for each and every river. Such as an ongoing commitment is operationalized by everyone, reflecting the ways we live with rivers.

1.2   Although Rivers Are Precious, They Are Treated with Disdain

As symbols of purity, renewal, timelessness, and healing, rivers have shaped human spirituality like few other features of the natural world. … Evoking magic, mystery, and beauty, rivers have inspired painters, poets, musicians, and artists of all kinds throughout history, adding immeasurably to the human experience. Sandra Postel and Brian Richter (2003, p. 6)

The interconnectedness of healthy river systems plays a critical role in sustaining life on Earth. These critical arteries and lifelines act as biological engines, connecting the webs of life of freshwater, terrestrial and marine environments. Access to water and fertile soils on floodplains supported the development of irrigated agriculture which underpinned the emergence of hydraulic civilizations and the world's first cities (Chap. 2). Indeed, humans have been acting as ecosystem engineers for thousands of years, fashioning increasingly domesticated ecosystems in many parts of the world (Kareiva et al. 2007). In the quest to provide resources and services to meet human needs, rivers have long been focal points of technical innovation, exemplified by devices to extract, transfer and store water and sophisticated engineering applications to build bridges, water treatment facilities and networks of navigable channels. Virtually all large rivers now support opportunities for transport, trade, renewable energy production and a host of other services of enormous benefit to society (see Box 1.1).

> **Box 1.1 The Role of Rivers in the Development of the United States of America**
> Rivers played a central role in the development of various political and socio-economic aspects of American life. Doyle (2018) shows how concerns for the management of river systems influenced issues such as federalism, taxation, regulation, conservation and resource use. Rivers fuelled early phases of the Industrial Revolution. Many

tens of thousands of mill dams were constructed along mid-Atlantic streams from the late seventeenth to the early twentieth century (Walter and Merritts 2008). For example, a 72 km stretch of the Blackstone River that extended from Worcester, Massachusetts, to Providence, Rhode Island, had over 1100 water mills at its peak.

Rivers and canals supported industrial expansion to the Midwest. Interstate river navigation supported by flood control and canalization programmes designed and implemented by the Army Corps of Engineers, helped to shape the roots of the U.S. Constitution. In a much less formal and planned manner, rivers played a central role in the California gold rush in 1848. The National Reclamation Act (1902) orchestrated under the presidency of Theodore Roosevelt viewed human control of rivers as fundamental to economic and social advancement (Postel and Richter 2003).

For the first 150 years or so of the nation, state-based federalism drove American lives, with central government playing a relatively limited role in day to day affairs. However, around the time of the Great Depression it became evident that the states did not have sufficient resources to build dams, run pipelines and develop power grids across the country, so major structural changes were implemented to apply broad-scale water resource development programmes. For example, the Tennessee Valley Authority was a key part of the New Deal. Dam and water transfer megaprojects transformed the American southwest. Levees now disconnect the Mississippi River from 90% of its floodplains (McPhee 1989).

Marked differences in legal frameworks underpin approaches to water management in differing jurisdictions of the United States (Postel and Richter 2003). Most eastern states apply the riparian doctrine of water law, according to which parties adjacent to rivers and streams can make reasonable use of those waters. Rather than individuals owning rights to water, the state permits the use of water bodies. This is often accompanied by conditions or requirements to ensure that such uses do not cause unreasonable harm to others. By contrast, most western states abide by the prior appropriation doctrine encapsulated by the mottos "first in time, first in right" and "use it or lose it". The legacy of these historical framings has a profound impact upon contemporary societal relations to rivers in differing parts of the country.

The use of rivers for drinking, navigation, power, waste disposal and sewage management have influenced concepts of regulation and ownership in the United States (Doyle 2018). Municipal bonds were created so that cities could borrow enough money to build water infrastructure and sewer systems. Taxation provided capital for government programmes to address river pollution and water treatment problems. Inevitably, as water came to be valued as a commodity that could be purchased, contentious disputes arose among private individuals, corporations and different levels of government, prompting ongoing legal deliberations regarding who controls or owns the land under rivers and the water that flows through them.

Many people have emotional, aesthetic and recreational connections to rivers, playing, fishing, swimming and boating in them, enjoying them for simply being as they are. Duncan (2001) highlights the tragedy of living lives that lack intimacy with rivers. Drew (2013) outlines how feelings of intimacy for the Ganga in India are associated with issues of love and loss and concerns for identity. As noted by Geoff Park (1995, p. 127): "And as the river landscape is filled with history, it is filled with emotion." Rivers are celebrated through various water rites, in art, dance, poetry and theatre (Ellis 2018). In a beautiful poem, Alice Oswald (2012) interweaves soundscapes of the River Dart in southwest England with the voices of human relations to the river. Many rivers have been worshipped as sacred entities for millennia, imbued with religious, spiritual and cultural meaning through various rituals and ceremonies. For many people, rivers are an important part of their identity, expressed through ancestral relations, nostalgic connections or formative life experiences (Palmer 1994). Authors such as Mark Twain, Joseph Conrad and Herman Hesse reflect upon various lessons in life learnt through explorations, encounters and spiritual quests along rivers. Songs like 'Take me to the river' or 'Waterfalls' evoke emotional connections to watercourses. The Portuguese poet, Fernando Pessoa (2002) expresses affective memories of Lisbon through a series of river stories in "The Book of Disquiet". Confronting realities of wounded waterscapes convey a profound sense of cultural and environmental loss (Tousley 2018).

Sometimes rivers bring people together, in other instances they keep them apart. As a result of historical imprints extending back to the days of the Roman Empire, the Danube River separates Catholicism and the Eastern Orthodox Church while the Rhine River demarcates the separation of Catholicism and Protestantism (Dartnell 2018). Some rivers have been used as agents of war, keeping invaders at bay. For example, floodwaters were routed through embankments along the lower course of the Yellow River in China during the Sino-Japanese conflict in 1938.

Prior to the nineteenth century, human impacts upon river systems were relatively localized, induced primarily by indirect responses to land use change. However, once the Industrial Revolution gathered momentum, systematic changes to rivers have taken place across the world (Best 2019; Macklin and Lewin 2019). The quest to make drylands wetter and wetlands drier has been accompanied by efforts to make river systems predictable and reliable. A command and control approach to river management tamed unruly rivers and harnessed their waters for irrigation, transportation, flood control and hydropower. The interests of hydraulic efficiency, navigability and flood mitigation created a uniformity of riverscapes and flows across quite different settings. Unfortunately, rivers and aquifers have been relentlessly over-exploited and freshwater no longer reaches the sea for months at a time along rivers such as the Colorado, Rio Grande, Yellow, Indus, Ganges, Amu Darya, Murray and Nile (Groenfeldt 2016). Rivers have been treated as gutters and drains for sewage and waste products. Although sewage treatment facilities and management of contaminants and toxic waste are conducted much more effectively than in previous eras, the legacy of pollutants from industrial, agricultural and mining activities continues to exert a significant impact upon water quality in many areas. Population growth and climate change present increasing pressure on available water resources, prompting disputes among rights-holders and stakeholders related to the water-energy-food nexus (e.g. Hussey and Pittock 2012) and a myriad of other demands. Contestations are especially pronounced when they are part of transboundary disputes (e.g. Shiva 2002).

Despite their fundamental importance to human wellbeing, many rivers are in a perilous state (e.g. Millennium Ecosystem Assessment 2005; Vörösmarty et al. 2010; WWF 2016). Impacts are especially severe upon freshwater fish, mussels, crayfish, birds, amphibians and many other animal and plant species. Iconic freshwater megafauna such as sturgeons, river dolphins, hippopotamus, crocodiles and turtles are threatened by overexploitation, increasingly fragmented rivers, habitat degradation, pollution and species invasion (Carrizo et al. 2017; He et al. 2017,

2019). The baiji—a freshwater dolphin previously described as the Goddess of the Yangtze River—is functionally dead.

Although moves towards an era of river repair have taken place in recent decades, such that images of rivers choked by toxic waste products or use of channels as drains and open sewers are increasingly distant memories in many parts of the world, elsewhere such realities continue to this day. It is increasingly clear that the technical skills that brought about the degradation of river systems will not, in their own right, provide the required approach to bring about the scope and scale of river repair that is necessary. A different way of thinking is needed.

1.3   Socionatures: River As a Resource or Rivers As Lifeblood?

What is life? Is it inseparable from Earth? At the most elemental level, we living beings are not even properly things, but rather processes … the electrochemical processes that are life are entirely consistent with an origin in Earth's crust—our very chemistry tells us that we are, in all probability, of it. Tim Flannery (2010, pp. 40–41)

As prophetically expressed by Heraclitus (c. 535–c. 475 BC), you can never step in the same river twice. A river never goes back exactly to what it was in quite the same way. At the same time, the person who is seeing the river is never exactly the same as they were previously. Societal values (what it is hoped will be seen) and the shifting baseline of expectation (what is expected based on experience) change over time.

Notions of nature are inherently value-laden, reflecting an imperfect mix of morality and commerce, aesthetics and need, stewardship and politics. As noted by Ginn and Demeritt (2009, p. 308), "nature is as much a concept as it is a biophysical reality. Far from being something located 'out there', nature is also something with us 'in here' …." We construct nature, and our relationships to it, in different ways. As Budiansky (1995, p. 98) asserts: "Nature has no eternal plan, no timeless purpose. It is ever changing …."

Several centuries of manipulating rivers have obscured societal awareness of dependence upon the services provided by healthy aquatic ecosystems. All too frequently, we fail to respect the blue arteries of the earth that sustain the planet's life-support system (Postel and Richter 2003). Salmond (2017, p. 315) notes that "… the most intransigent obstacles to solving environmental (and other) challenges lie at the level of presupposition. For waterways, the assumption that they were created to serve human purposes drives towards their degradation and destruction."

As human society has transitioned from agricultural to increasingly urban lifestyles, with technology playing an ever-greater role in day to day lives, working relations to ecosystems and the rhythms of the natural world have been progressively eroded. We buy food and clothing in stores, rather than growing or making our own. We depend on technology to deliver water and energy. Divorcing ourselves from nature creates matter out of place (Muru-Lanning 2016), disconnecting society from metaphysical questions of identity and meaning.

Across much of the planet, rivers have been marginalized from the lives of many people—they have been 'othered'. Many urban streams have been 'buried alive' in underground pipes. Elsewhere, channelized and canalized are bereft of life, acting as 'zombie' forms, behaving like the living dead as clean water flows through a sterile but tidy channel. Countless other waterways continue to be treated with disdain as gutters, drains and sewers. Such actions separate society from living rivers in physical and psychological terms, unconsciously furthering neglect through disconnection from the values of healthy aquatic ecosystems (see Box 1.2).

Box 1.2 The Meanings of Water
Material flows of water are related to social flows of capital and power, as water becomes embodied in channels and networks that course across the landscape (Wilder and Ingram 2018). Modern water is characterized by its physical containment and isolation from people, wherein engineering measures maintain provision of supply, at a cost, allowing most of society to not have to think much about it (Linton 2010). Societal expectations have adjusted to norms in which infrastructure delivers water on-demand, through a tap. Such end-of-pipe framings disconnect society from cultural and environmental meanings of water.

However, water is much more than a utilitarian resource. Water and life are mutually constitutive. Meanings of water are physical and metaphorical, cultural and psychological, emotional and aesthetic (e.g. Bouleau 2014; Linton 2010; Strang 2004). As water circulates through social and environmental space, it remakes human and non-human relations (Linton and Budds 2014). Freshwater places are distinct, not abstract (Gibbs 2014).

The simple assertion that access to water of an appropriate quantity and quality is a basic human right privileges anthropocentric values over environmental values, the rights of the non-human world, and various non-use values such as symbolic, emotional, spiritual and religious connections to water—its role as lifeblood that is vital to life and wellbeing (Yates et al. 2017). If equity is to be served, water must be viewed as a social actor—a socio-natural resource, rather than merely a physical or natural resource. In a sense, water-as-a-resource treats water within a curative ethos, while water-as-lifeblood applies a preventative ethos including human and non-human actors that make up a waterscape (Yates et al. 2017).

Simply asking "What is a healthy river/society?" is an expression of concern for the world. Notions of healthy rivers reflect societal values, perceptions, attitudes and aesthetics (Table 1.1). Very different realities are engendered if society is viewed as part of a greater organism that is respected and nourished, relative to conceptualizations of nature as a machine that is designed to serve the human race (Taylor et al. 2016). Are humans viewed to be separate from nature, exploiting and controlling rivers to meet human needs, suppressing and subduing their variability as life is squeezed out of them? Or, are humans viewed as part of nature, living harmoniously with the inherent diversity, dynamics and evolutionary traits of living rivers? Are rivers embraced as focal points of societies, or are they seen as hazards that are feared by society, such that it turns its back to them (Box 1.3)? Is there an aesthetic preference for complex, messy channels or uniformly clean and tidy rivers? Is nurturing and looking after rivers an issue of personal responsibility, or is it merely a problem to be addressed by river managers?

Table 1.1 River values

Values	Examples
Economic and amenity values—technocentric relations of a command and control world	• Water supply for irrigated food production, industrial uses (including cooling water), domestic water supply and other consumptive uses • Hydropower • Navigation—channelization • Flood control • Resource use, such as sand/gravel extraction • Food, including customary harvest • Tourism activities • Ecosystem services

Environmental values—ecosystem functionality and the health of a living river	• Water quantity—the flow regime
	• Water quality
	• Sediment regime, hydraulics, channel adjustment, floodplain processes (alluvial soil formation and regeneration)
	• Physical habitat (geodiversity)
	• Biodiversity—aquatic and terrestrial life, connectivity relations
	  – Life cycles, food webs, predator-prey relations, species assemblages
	  – Endangered or threatened species; unique or rare attributes; invasive/exotic species; iconic (keystone) species
	  – Measures of ecosystem functionality
	• Biogeochemical processes
	• Interplay between abiotic and biotic attributes and processes; ecosystem engineers

Socio-cultural values—meanings of rivers	• Religious and spiritual values/connections • Option, bequest and intrinsic values • Physical health/wellbeing (polluted/contaminated water; risk of disease) • Mental wellbeing (aesthetics, artistic endeavours, psychological and therapeutic values, emotional connections) • Recreational values and sporting activities
Rights of the River	• The right to act as a living, emergent entity

Box 1.3 Perceptions of Floods
Value systems influence perceptions of floods. From an economic perspective they may be viewed as deadly, disruptive and expensive hazards—something to be managed out of existence. From an environmental perspective, floods are an inherent part of the range of variability of a river system, triggering for key ecological responses and driving morphological adjustments, nourishing and replenishing ecosystem values and facilitating processes of river recovery. From a socio-cultural and psychological perspective, floods provide a graphic reminder of the living entity with which societies live—something that is much bigger than ourselves, something that cannot be controlled. Flood problems are entirely anthropocentric concerns (Cioc 2002). Ironically, the more protection from floods that is provided and promised, the greater the costs to maintain measures that have been put in place to protect society and the greater the potential consequences of a disaster once barriers are breached. The logics of risk and impact are quite different when viewed from a community perspective, concerns of the (re)insurance industry, or the rights of a living river. Despite growing concerns for flood risk, more people live on floodplains than at any period in Earth history.

Viewed from a socio-economic perspective, rivers are conceptualized as machines to be manipulated to provide services and resources for the benefit of society (Nestler et al. 2012). A technocentric lens applies a command and control approach to assert human dominion and authority over a river, framing relations to nature as a form of master-servant relationship. Engineering and domesticated riverscapes with predictable, uniform, stable and hydraulically efficient channels, or flow that is buried within pipes, are manifestations of such a worldview.

Environmental values appraised through a scientific lens conceptualize rivers in two quite different ways. First, biodiversity and ecosystem values underpin assertions of a living river, framing notions of river health in relation to ecological principles such as ecosystem vigour, organization and resilience (Everard and Powell 2002; Karr 1999; Norris and Thoms 1999; Rapport et al. 1998). Second, an anthropocentric perspective views ecosystem services as forms of natural capital provided by the river that can be managed for the benefit of society. Examples include clean water to drink and fish to eat, moderating functions of floods and droughts, water purification/treatment, maintaining food webs, biodiversity conservation, delivering nutrients to estuaries and replenishing floodplain soils (Postel and Richter 2003).

Socio-cultural values emphasize concerns for wellbeing, encompassing a range of things that society cares about, including psychological, emotional and relational aspects of rivers. A socio-ecological lens conceptualizes mutual interactions between society and the river, wherein riverscapes not only have value, they have meaning. Sometimes ancestral relations include more-than-human associations in which rivers are enduring reflections of cultural values, past and present (Nassauer 1995; Thomas 2015). *Finding the Voice of the River* builds upon these relationships, incorporating assertions of river rights that protect the values of the river as a living, emergent entity.

1.4   A Living River Ethos

When we view the world and our place within it we need to understand that we share a common history with other living beings. We should realize that we are part of a very diverse whole, both in terms of life forms as well as in ways of doing and being. If we stop thinking of humans as being separate from nature, it then becomes possible to suggest models of conduct where biodiversity and cultural diversity are respected and interlinked. Sign in La Musee des Confluences, Lyon, France

Rivers express themselves in different ways as they adjust to prevailing conditions (Box 1.4). A gorge may have an imposing presence, firmly etched into bedrock. Discontinuous watercourses may be subtle and ambiguous, struggling to flow. Rivers have various moods as they adjust and evolve. Trajectories and rates of change are not always predictable, as complexities and contingencies underpin emergent properties and inherent uncertainties of living systems (Hillman and Brierley 2008).

Box 1.4 Act Me a River …

Building upon an exercise outlined by Fisher (1997), I run an interactive class that asks students to "Draw me a river". Are rivers conveyed using a 2D or a 3D representation—a cross-section or planform perspective, or a block diagram? What type of river is drawn? Are channel and floodplain compartments included? Is it a discontinuous watercourse, or does as aesthetic preference for meandering channels come to the fore (Kondolf 2006)? Are vegetation, fish and anthropogenic structures (dams, levees, etc.) included?

Extending such perspectives, I start my upper level rivers class by asking students to form groups and "Act me a river", with other groups guessing the type of river that is being performed through expressions of behavioural adjustments. Some exceptionally creative, energetic and evocative displays of river dynamics have been presented. Rumbling boulders in a steep headwater stream conveyed by a group of gymnastic parkour practitioners was especially memorable.

Approaches to 'living with living rivers' have been practised as core elements of traditional cultures and societies for many thousands of years, encapsulated within entangled notions of nature, society and place (Berkes 1999; Davis 2009; Smith 2017). Through such lenses, responsible stewardship frames humans as partners rather than proprietors, custodians and guardians rather than owners, recognizing the interconnectedness of all that exists. This entails respect for what is gone before and what will come into the future, embracing ancestral relations wherein the dead walk with the living, protecting from beyond the grave (Langton 2002). Reciprocal and relational framings recognize that what we do to the land, we do to ourselves. Such understandings prompt efforts to look after the land as the land looks after us. Approaches to biocultural restoration recognize that sustaining the land sustains ways of living. In this light, healing the land supports societal healing (Blackstock 2001; Kimmerer 2011). Cree elders cited by LaBoucane-Benson et al. (2012, p. 7) assert that: "We are the water, and the water is us". As the arteries of the land, rivers simultaneously flow through human veins as life-giving forces (Fox et al. 2017). In such conceptualizations, the environment is sacred—an organic and emergent organism—a sentient entity. A deep respect for country emerges from such framings (Box 1.5).

Box 1.5 Listening to Country ...
In his keynote address at a conference in Auckland in July 2018, Richie Howitt outlined Aboriginal perspectives upon living with and as a part of the land in Australia, expressed as 'Listening to Country'. The land of the people is the people themselves and, in turn, the people are the human face of the land (Wright 2015, p. 403). In acknowledging respect for country, the land and its rivers have a voice ...

In her wanderings with the Ngardi and Kukatja peoples of the Tanami Desert in the East Kimberley of Western Australia, Kim Mahood (2016, p. 43) reflects upon indigenous relations with the land, with country, with place, reflecting upon "walking in the shared mindscape of the women who have brought me here. They are themselves in the here and now, and they are the sisters (hills) that exist as features of the landscape; simultaneously individual modern beings and the embodiment of ancient collective presences manifest as the sacred geography of their world." Later in her book, Mahood (2016, p. 231) comments that particular connections to a place may die out with the loss of indigenous languages, such that conversations with ancestors fall silent.

Australian researchers working on intersecting relationships between land and people publish their findings as Bawaka Country, a complex relational and sentient entity (e.g. Country et al. 2019). Songspirals bring together diverse more-than-human worlds and foster co-becoming, wherein an emergent reality and an ancestral reality are one and the same. As outlined by Country et al. (2019, p. 2): "Songspirals are a keening and singing of, with, by, for and *as* Country" (italics in original). Personification of landscapes as authors does not fit elegantly into the world of Search Engine citation metrics and h-factors.

Recent technological innovations have developed a range of sensory devices that use the voice of the river to assess various attributes of river behaviour. Acoustic profiling techniques are used to predict sediment flux (e.g. Geay et al. 2017; Lumsdon et al. 2018). Passive acoustic bioacoustics and automated monitoring techniques support readily accessible and non-invasive approaches to 'river listening' as bird and frog calls provide measures of wetland health to support conservation programmes (e.g. Barclay et al. 2018; Linke et al. 2018). Lockwood (2007) uses sound maps to appraise sensory relationships to rivers.

Much of the thinking in this book is inspired by Māori relations to rivers—ways of being expressed by the indigenous peoples of Aotearoa New Zealand (Hikuroa et al. 2019). In *Te Ao Māori* (the Māori world), people are not only 'of the land' but 'as the land' (Te Aho 2010). Ancestors are literally planted in the earth, known as *tāngata whenua* (land people). In the ontologic of Māori language, people, land and ancestors are literally 'the same thing' (Salmond 2014). Past, present and future are with us at all times, as the dead walk with the living, protecting from beyond the grave, acting as custodians and guardians of the river (*kaikiakitanga*), not owners. Harming a river inflicts harm upon ancestors. A river is not just a resource to be used or a hazard to be controlled, it is an ancestral force to be lived with, reckoned with, and respected: As rivers nurture us, we have a responsibility to nurture them (Te Aho 2010).

Viewed through a Māori lens, the economy is a wholly owned subsidiary of the environment. An *awa* is not just a river—it is an interconnected, living system (Salmond 2014). Rivers have their own life force (*mauri*), their own spiritual energy and their own powerful identities (Te Aho 2010). Embracing Māori thinking about rivers as living beings sees people as part of the system, without arrogance or assumptions that they are in control of it (Salmond 2017). Notions of *ora* (health, wellbeing) encapsulate a state of peace, prosperity and wellbeing for people, plants and animals, as well as the river. *Ora* is not simply a biophysical or even socio-ecological concept; it has philosophical (ontological), political and spiritual dimensions (Salmond 2014, 2017). Its contrary, *mate*, refers to a state of ill-health or dysfunction, as a result of faltering or failing interdependencies within a system. What constitutes a state of *ora*, and what it takes to look after it, reflect river-specific attributes, values, relationships, dynamics and evolutionary traits.

Herein lies the basis to live respectfully with rivers, working with the river rather than against it. If the *mauri* of a river is not respected, or if people attempt to assert dominance over it, it loses its vitality and force, and those who depend upon it will ultimately suffer.

1.5   Engaging with the More-than-Human World: What Does It Mean to Find the Voice of the River?

To know the geography of a place is to know why we have always made stories in which our own human stuff is indivisible from the stones and creeks and hills and growing things. Kim Mahood (2016, p. 2)

Relations to the more-than-human world expressed through *Finding the Voice of the River* can be conceptualized as a form of biophilia, a psychological, often subconscious orientation and connection of being attracted to all that is alive and vital (Fromm 1964; Wilson 1984). After all, a love of life and a love of rivers help to sustain us (cf., Fausch 2015). *Finding the Voice of the River* embraces sentient relationships with the more-than-human world, paralleling notions of thinking like a mountain (Brower and Chapple 1995; Freudenburg et al. 1995; Leopold 1949), listening to glaciers (Cruikshank 2014) and working with the organismic properties of rivers (Leopold 1994) and the sensory capacities of plants (Chamovitz 2012). As envisaged within the organic architecture of Frank Lloyd Wright, endeavours set out to be not merely 'at one' with the landscape, but 'of' the landscape.

Extending notions of a culture and duty of care to non-humans as well as fellow humans recognizes that "we live in a moral landscape governed by relationships of mutual responsibility which are simultaneously material and spiritual" (Kimmerer 2011, p. 268). If we are part of the river and the river is part of us, it is subject to sensations, it experiences responses, it has a form of consciousness. In other words, it is sentient. Singer (1972, p. 231) highlights a moral imperative to treat sentient beings in ways that avoid suffering and maximize wellbeing, noting: "If it is in our power to prevent something bad from happening, without thereby sacrificing anything of comparable moral importance, then we ought, morally, to do it." Just as the animal rights movement makes an ever-growing impact in societal relations to the more-than-human world, this book strives to find and express the voice of rivers as a living entity. Conceiving a river as sentient draws attention not only to intrinsic values, but also to concerns for wellbeing—both physical and mental; how the river works and its various moods and behaviours. All too often, we fail to appreciate the virtues of freedom until it is taken away from us. Liberally paraphrasing Jean-Jacques Rousseau (1712–1778), rivers may have been born free, but most everywhere, their activities have been restrained. Inevitably, nature fights back (see Box 1.6).

Box 1.6 Silenced and Zombified Rivers Express Their Voice
River channels that are fixed in place and locked in time are stifled. They are physically and psychologically dead, acting like zombie rivers (after Gleick et al. 2014)! Over-engineered river have been silenced; their voice has been taken away (McCully 1996). If a river's flow is its heartbeat, a resuscitator is required.

Restless rivers strive to break free from restraints, taking advantage of any opportunity to fight back. In 1997, parts of the Bečva River, a tributary of the Morava River in the Czech Republic, broke free from their stopbanks during a flood. Given a lack of resources, engineering works were not reapplied and the channel was allowed to 'renaturalize'. The river rediscovered its voice. Erosional and depositional processes regenerated the river. Water chemistry and the thermal regime improved. Habitat diversity and the range of flora and fauna increased. Local communities reconnected to a living river as a place to fish and play.

Acting consciously, minimizing harm and allowing rivers to express their voice provides a unifying platform for collective efforts to listen to and live generatively with each and every river. Prospectively, thinking and acting on behalf of the river presents a disarming and upbeat platform for collective engagement. Why would anyone, knowingly, want to harm a river? If a particular action harms the river, don't do it!

Three key issues set the relational approach to living with rivers advocated in this book apart from contemporary approaches to river management:

(a) A non-materialist philosophy of nature emphasizes the inextricability of social and natural objects. A more-than-human lens focusses attention upon relations to rivers not only as determinants of river health, but also the collective wellbeing of society. Taking care for relationships takes care of life itself.
(b) Living with rivers reflects concerns for all rivers, everywhere, while respecting the place-based values of each river system.
(c) Relational connections to a culture and duty of care engender ongoing commitment to the river and each other rather than conceiving such things as a fix and forget project.

Such framings underpinned the granting of rights to the Whanganui River in Aotearoa New Zealand as a living entity in 2017 (Brierley et al. 2018; Charpleix 2018; Ruru 2018).

1.6   THE RIGHTS OF THE RIVER

The health of rivers and therefore the well-being of life on earth demands a new commitment of heart and soul that leads to positive reform at the individual, community, and national levels. Tim Palmer (1994, pp. 223–224)

The notion of water as a public good can be traced back to the public trust doctrine that emerged through the Institutes of Justinian as part of Roman civil law dating back to A.D. 530. It states: "By the law of nature, these things are common to mankind—the air, running water, the sea, and consequently the shores of the sea" (quoted in Postel and Richter 2003, p. 84). As legal framings are recognized by all parties, this is the space where transformations in practice come into effect. Prospectively, reconceptualizations of relationships to the more-than-human world could open up debates on economic, cultural and political issues of inter-dependence, justice and rights. What are the moral and ethical obligations of society towards the environment? Is it time to draft a new Magna Carta, establishing that no human is above nature, creating an eco-constitutional state in which principles of Earth Jurisprudence move from an Anthropocentric ontology to a more Earth-centred one (Bosselmann 2008)?

Rights of Nature is grounded in the recognition that humankind and Nature share a fundamental, non-anthropocentric relationship given our shared existence on this planet (Kothari and Bajpai 2017). Rather than being regarded merely as property or a commodity under the law, nature, like humans, is recognized as a rights-bearing entity, with rights that need to be protected (Stone 1973). And we—the people—have the legal authority and responsibility to enforce these inalienable rights on behalf of ecosystems.

An international chorus of legal and constitutional changes assert the rights of nature, including initiatives in Bolivia, Colombia, Ecuador, India and New Zealand, among others (Boyd 2017). Ecuador was the first country in the world to codify the Rights of Nature in their Constitution in 2008. Article 71 of the Constitution states: "Nature or 'Pachamama', where life is reproduced and exists, has the right to exist, persist, maintain and regenerate its vital cycles, structure, functions and evolutionary processes" (Rühs and Jones 2016).

Rights of the River have played a central role in assertions of the rights of nature (e.g. Chapron et al. 2019; Charpleix 2018; Kinkaid 2019; O'Donnell 2018; Wilson and Lee 2019). In March 2017, the Whanganui

River on the North Island of New Zealand was granted the status of legal personhood. A new legal entity was created, *Te Awa Tupua*, referring to "an indivisible and living whole from the mountains to the sea, incorporating the Whanganui River and all of its physical and metaphysical elements" (*Te Awa Tupua* [Whanganui River Claims Settlement] Act, section 13(b)). This law conferred *Te Awa Tupua* "all the rights, powers, duties, and liabilities of a legal person" (section 14(1)), to be expressed through a newly created governance authority.

Granting rights to the Whanganui River takes us beyond a perspective in which "Nature's out there and we're over here". As noted by Morris and Ruru (2010, p. 58), "The beauty of the concept is that it takes a western legal precedent and gives life to a river that better aligns with a Māori worldview that has always regarded rivers as containing their own distinct life forces." As a sentient multidimensional entity, the Whanganui River has agency, law and spirit in its own right (Box 1.7).

Box 1.7 Two Degrees of Separation ...

Aotearoa is a pretty small and friendly place. We often refer to 'two degrees of separation' in terms of who knows who. In a recent lecture on Māori relations to rivers, and the rights of the Whanganui River, I included a quote from a Māori elder: Rangiwaiata Rangitihi Tahuparae "*Ko au te awa, ko te awa ko au*". Suddenly one of the students in the class perked up: "That's my dad". The actual proverb (*whakatauki*) is:

> *E rere kau mai te awa nui nei.*
> *Mai i te kāhui maunga ki Tangaroa.*
> *Ko au te awa.*
> *Ko te awa ko au.*
> *This great river flows from the mountains to the sea*
> *I am the river, and the river is me.*

Building on connections to her ancestral river, Nga Remu Huia Tahuparae offers the following thoughts on legal personhood of the Whanganui River.

What does legal personhood mean to me? I see it as a vehicle to transform people's *whakaaro* (ways of thinking) and relationship to the *awa* (river), helping to understand what is needed to respect and understand the *awa's* needs as it flows from the mountains to the sea. It's a device to shift the *whakaaro* of those in power and those who continue to take and destroy what the *awa* needs to flourish.

Te Awa Tupua kawa (reframings under the Whanganui River Claims Settlement Act) presents a framework that can spread throughout Aotearoa, reminding other guardians of the rivers what is possible. Under *Te Tiriti o Waitangi*, all rivers in Aotearoa New Zealand have the right to flow free and be in a state of *ora* (wellbeing).

It is essential to return to a knowledge base in which the seen and unseen are viewed collectively when making decisions regarding any force of nature. Disrepair will continue so long as we separate the physical and spiritual aspects of the river. Legal personhood supports the integration of *mātauranga Māori* and western science, working together to heal rivers, revitalizing the *mouri* of the *awa* (taking steps to enhance the life force of the river). In many ways, *mouri* is innate, intuitive—sensory relations and connections that cannot necessarily be rationalized, yet they are so very real.

The *awa* is a living, powerful and beautiful being. It has existed in geologic timeframes while we have only just arrived. It is our lifeline. It is not something that belongs to us; rather, it is something to which we belong; as such, it is our duty to protect it.

When working with a river, the first place to start is to talk to it, finding out who it is and what it needs to flow in a state of *ora* (health). The following *whakatauki* presents an account of relations to the river:

> *Kauaka e kōrero mō te awa, engari kōrero ki te awa.*
> *Don't merely talk about the river, rather speak and commune with the river.*
> ~ *Te Tira Hoe Waka o Whanganui*

Rights provide a legal stick to reframe policy and governance arrangements around a different set of values that allows the voice of the river to emerge. For the river to have rights in the eyes of law would mean that a suit could be brought in the name of the river, injury can be recognized, the polluter can be held liable for harming, and compensation will be paid that would benefit the river (Kothari and Bajpai 2017). Assertions of river rights recognize the inherent rights of rivers to exist regardless of their economic value.

Inevitably, conflict abounds in negotiating the interests of rivers as public spaces of communal activity and engagement, relative to more regulated places of ownership, abstraction and supply (Dudley 2017). Several key questions are yet to be answered (based on Kothari and Bajpai 2017; Rühs and Jones 2016):

- What does it mean for a river to have rights? Do rights just include the river, or also the species that inhabit and use it? What does the river want?
- How can a river, with no voice of its own, ensure that its rights are met? Who speaks for the river? How do we decide who or what is an appropriate guardian? If rights are violated, who asks for compensatory action? Who would be the beneficiary of such compensation? What happens if the parents or custodians fail to discharge their duty?
- What does it mean for a river to have rights as a "person"? Only humans can be moral agents and nature is incapable of claiming rights of its own, so should rights of nature be a subject of law? However, newborns have rights, even though they cannot distinguish between right and wrong (Rühs and Jones 2016). A human right comes with the possibility or promise of restitution, redressal and compensation in the event of violations of such a right. What would this mean for a river? In the case of violation of the right, what will count as damage?
- Can a river's right to flow free be read as equivalent to a person's fundamental right to speech and expression?

Prospectively, a 'rights' framing represents a step along the way to realizing that rivers, the sea, forests and the earth are more powerful and ancient than people; that humanity's presence on Earth is contingent on realizing healthy relationships with other living beings and systems (Salmond 2017). Ultimately, relationships to rivers are an embodiment of relationships to each other.

REFERENCES

Barclay, L., Gifford, T., & Linke, S. (2018). River listening: Acoustic ecology and aquatic bioacoustics. *Leonardo, 51*(3), 298–299.

Berkes, F. (1999). *Sacred Ecology. Traditional Ecological Knowledge and Resource Management.* London: Routledge.

Best, J. (2019). Anthropogenic stresses on the world's big rivers. *Nature Geoscience, 12,* 7–21.

Blackstock, M. (2001). Water: A first nations' spiritual and ecological perspective. *BC Journal of Ecosystems and Management, 1,* 2–14.

Bosselmann, K. (2008). *The Principle of Sustainability: Transforming Law and Governance.* London: Routledge.

Bouleau, G. (2014). The co-production of science and waterscapes: The case of the Seine and the Rhône Rivers, France. *Geoforum, 57,* 248–257.

Boyd, D. R. (2017). *The Rights of Nature.* Toronto, ON: ECW Press.

Brierley, G. J., & Fryirs, K. A. (Eds.). (2008). *River Futures.* Washington, DC: Island Press.

Brierley, G., Tadaki, M., Hikuroa, D., Blue, B., Šunde, C., Tunnicliffe, J., & Salmond, A. (2018). A geomorphic perspective on the rights of the river in Aotearoa New Zealand. *River Research and Applications.* https://doi.org/10.1002/rra.3343

Brower, D., & Chapple, S. (1995). *Let the Mountains Speak, Let the Rivers Run: A Call to Those Who Would Save the Earth.* New York: Harper Collins.

Budiansky, S. (1995). *Nature's Keepers. The New Science of Nature Management.* London: Weidenfeld & Nicolson.

Capra, F. (1982). *The Turning Point. Science, Society and the Rising Culture.* New York: Simon and Schuster (Bantam Paperback, 1983).

Carrizo, S. F., Jähnig, S. C., Bremerich, V., Freyhof, J., Harrison, I., He, F., ... Darwall, W. (2017). Freshwater megafauna: Flagships for freshwater biodiversity under threat. *BioScience, 67*(10), 919–927.

Chamovitz, D. (2012). *What a Plant Knows: A Field Guide to the Senses.* New York: Scientific American/Farrar, Straus and Giroux.

Chapron, G., Epstein, Y., & López-Bao, J. V. (2019). A rights revolution for nature. *Science, 363*(6434), 1392–1393.

Charpleix, L. (2018). The Whanganui River as Te Awa Tupua: Place-based law in a legally pluralistic society. *The Geographical Journal, 184*(1), 19–30.

Cioc, M. (2002). *The Rhine: An Eco-Biography.* Seattle, WA: University of Washington Press.

Country, B., Suchet-Pearson, S., Wright, S., Lloyd, K., Tofa, M., Burarrwanga, L., ... Maymuru, D. (2019). Bunbum ga dhä-yu t agum: To make it right again, to remake. *Social & Cultural Geography.* https://doi.org/10.1080/14649365.2019.1584825

Cruikshank, J. (2014). *Do Glaciers Listen? Local Knowledge, Colonial Encounters, and Social Imagination.* Vancouver, BC: UBC Press.

Dartnell, L. (2018). *Origins. How the Earth Made Us.* London: The Bodley Head.

Davis, W. (2009). *The Wayfinders: Why Ancient Wisdom Matters in the Modern World.* House of Anansi, Toronto, Canada.

Doyle, M. (2018). *The Source: How Rivers Made America and America Remade Its Rivers.* New York: Norton.

Drew, G. (2013). Why wouldn't we cry? Love and loss along a river in decline. *Emotion, Space and Society, 6,* 25–32.

Dudley, M. (2017). Muddying the waters: Recreational conflict and rights of use of British rivers. *Water History, 9*(3), 259–277.

Dufour, S., & Piégay, H. (2009). From the myth of a lost paradise to targeted river restoration: Forget natural references and focus on human benefits. *River Research and Applications, 25*(5), 568–581.

Duncan, D. J. (2001). *My Story as Told by Water.* San Francisco: Sierra Book Clubs.

Ellis, J. (Ed.). (2018). *Water Rites: Reimagining Water in the West.* Calgary: Calgary University Press.

Everard, M., & Powell, A. (2002). Rivers as living systems. *Aquatic Conservation: Marine and Freshwater Ecosystems, 12*(4), 329–337.

Fausch, K. D. (2015). *For the Love of Rivers. A Scientist's Journey.* Corvallis: Oregon State University Press.

Fisher, S. G. (1997). Creativity, idea generation, and the functional morphology of streams. *Journal of the North American Benthological Society, 16*(2), 305–318.

Flannery, T. F. (2010). *Here on Earth. A Natural History of the Planet.* New York: Atlantic Monthly Press.

Fox, C. A., Reo, N. J., Turner, D. A., Cook, J., Dituri, F., Fessell, B., … Turner, A. (2017). "The river is us; the river is in our veins": Re-defining river restoration in three Indigenous communities. *Sustainability Science, 12*(4), 521–533.

Freudenburg, W. R., Frickel, S., & Gramling, R. (1995). Beyond the nature/society divide: Learning to think about a mountain. *Sociological Forum, 10*(3), 361–392.

Friends of the Earth. (1972). *Only One Earth. The Care and Maintenance of a Small Planet.* Harmondsworth: Penguin.

Fromm, E. (1964). *The Heart of Man.* New York: Harper & Row.

Geay, T., Belleudy, P., Gervaise, C., Habersack, H., Aigner, J., Kreisler, A., … Laronne, J. B. (2017). Passive acoustic monitoring of bed load discharge in a large gravel bed river. *Journal of Geophysical Research: Earth Surface, 122*(2), 528–545.

Gibbs, L. (2014). Freshwater geographies? Place, matter, practice, hope. *New Zealand Geographer, 70*(1), 56–60.

Ginn, F., & Demeritt, D. (2009). Nature: A contested concept. In N. J. Clifford, S. L. Holloway, S. P. Rice, & G. Valentine (Eds.), *Key Concepts in Geography* (2nd ed., pp. 300–311). Los Angeles: Sage.

Gleick, P. H., Heberger, M., & Donnelly, K. (2014). Zombie water projects. In P. H. Gleick (Ed.), *The World's Water* (pp. 123–146). Washington, DC: Island Press.

Groenfeldt, D. (2016). Cultural water wars: Power and hegemony in the semiotics of water. In C. M. Ashcraft & T. Mayer (Eds.), *The Politics of Fresh Water. Access, Conflict and Identity* (pp. 143–156). London: Routledge.

He, F., Zarfl, C., Bremerich, V., David, J. N., Hogan, Z., Kalinkat, G., … Jähnig, S. C. (2019). The global decline of freshwater megafauna. *Global Change Biology.* https://doi.org/10.1111/gcb.14753

He, F., Zarfl, C., Bremerich, V., Henshaw, A., Darwall, W., Tockner, K., & Jähnig, S. C. (2017). Disappearing giants: A review of threats to freshwater megafauna. *Wiley Interdisciplinary Reviews: Water, 4*(3), e1208.

Hesse, H. (1951). *Siddhartha.* New York: Bantam.

Hikuroa, D., Brierley, G. J., Tadaki, M., Blue, B., & Salmond, A. (2019). Restoring socio-cultural relationships with rivers: Experiments in fluvial pluralism from Aotearoa New Zealand. In Cottet, M., Morandi, B, & Piégay, H. (Eds). Socio-Cultural Perspectives in River Restoration. Wiley (in press).

Hillman, M., & Brierley, G. J. (2008). Restoring uncertainty: translating science into management practice. In G. J. Brierley & K. A. Fryirs (Eds.), *River Futures: An Integrative Scientific Approach to River Repair* (pp. 257–272). Washington, DC: Island Press.

Hussey, K., & Pittock, J. (2012). The energy–water nexus: Managing the links between energy and water for a sustainable future. *Ecology and Society, 17*(1).

Kareiva, P., Watts, S., McDonald, R., & Boucher, T. (2007). Domesticated nature: Shaping landscapes and ecosystems for human welfare. *Science, 316*(5833), 1866–1869.

Karr, J. R. (1999). Defining and measuring river health. *Freshwater Biology, 41*(2), 221–234.

Kimmerer, R. (2011). Restoration and reciprocity: The contributions of traditional ecological knowledge. In D. Egan, E. E. Hjerpe, & J. Abrams (Eds.), *Human Dimensions of Ecological Restoration* (pp. 257–276). Washington, DC: Island Press.

Kinkaid, E. (2019). "Rights of nature" in translation: Assemblage geographies, boundary objects, and translocal social movements. *Transactions of the Institute of British Geographers.* https://doi.org/10.1111/tran.12303

Kondolf, G. M. (2006). River restoration and meanders. *Ecology and Society, 11*(2).

Kothari, A., & Bajpai, S. (2017). We are the river, the river is us. *Economic and Political Weekly, 52,* 103–109.

LaBoucane-Benson, P., Gibson, G., Benson, A., & Miller, G. (2012). Are we seeking Pimatisiwin or creating Pomewin? Implications for water policy. *International Indigenous Policy Journal, 3*(3). https://doi.org/10.18584/iipj.2012.3.3.10. Retrieved from https://ir.lib.uwo.ca/iipj/vol3/iss3/10.

Langton, M. (2002). The edge of the sacred, the edge of death: Sensual inscriptions. In B. David & M. Wilson (Eds.), *Inscribed Landscapes: Marking and Making Place* (pp. 253–269). Honolulu: University of Hawaii Press.

Leopold, A. (1949). *A Sand County Almanac and Sketches Here and There.* Oxford: Oxford University Press.

Leopold, L. B. (1994). River morphology as an analog to Darwin's theory of natural selection. *Proceedings of the American Philosophical Society, 138*(1), 31–47.

Linke, S., Gifford, T., Desjonquères, C., Tonolla, D., Aubin, T., Barclay, L., … Sueur, J. (2018). Freshwater ecoacoustics as a tool for continuous ecosystem monitoring. *Frontiers in Ecology and the Environment, 16*(4), 231–238.

Linton, J. (2010). *What Is Water?: The History of a Modern Abstraction.* Vancouver, BC: UBC Press.

Linton, J., & Budds, J. (2014). The hydrosocial cycle: Defining and mobilizing a relational-dialectical approach to water. *Geoforum, 57,* 170–180.

Lockwood, A. (2007). What is a river. *Soundscape: The Journal of Acoustic Ecology, 7*(1), 43–44.

Lumsdon, A. E., Artamonov, I., Bruno, M. C., Righetti, M., Tockner, K., Tonolla, D., & Zarfl, C. (2018). Soundpeaking–Hydropeaking induced changes in river soundscapes. *River Research and Applications, 34*(1), 3–12.

Macklin, M. G., & Lewin, J. (2019). River stresses in anthropogenic times: Large-scale global patterns and extended environmental timelines. *Progress in Physical Geography: Earth and Environment, 43,* 3–23.

Mahood, K. (2016). *Position Doubtful. Mapping Landscapes and Memories.* Melbourne, VIC: Scribe.

McCully, P. (1996). *Silenced Rivers: The Ecology and Politics of Large Dams.* London: Zed Books.

McPhee, J. (1989). *The Control of Nature.* New York: Farrar, Straus & Giroux.

Meadows, D. H., Meadows, D. H., Randers, J., & Behrens, W. W., III. (1972). *The Limits to Growth: A Report for the Club of Rome's Project on the Predicament of Mankind.* New York: Universe Books.

Millennium Ecosystem Assessment. (2005). *Ecosystems and Human Well-Being.* Washington, DC: Island Press.

Monbiot, G. (2019, April 15). Only rebellion will prevent an ecological collapse. *The Guardian.*

Morris, J. D., & Ruru, J. (2010). Giving voice to rivers: Legal personality as a vehicle for recognising Indigenous peoples' relationships to water? *Australian Indigenous Law Review, 14*(2), 49–62.

Muru-Lanning, M. (2016). *Tupuna Awa. People and Politics of the Waikato River* (p. 230). Auckland: University of Auckland Press.

Nassauer, J. I. (1995). Culture and changing landscape structure. *Landscape Ecology, 10*(4), 229–237.

Nestler, J. M., Pompeu, P. S., Goodwin, R. A., Smith, D. L., Silva, L. G., Baigun, C. R., & Oldani, N. O. (2012). The river machine: A template for fish movement and habitat, fluvial geomorphology, fluid dynamics and biogeochemical cycling. *River Research and Applications, 28*(4), 490–503.

Norris, R. H., & Thoms, M. C. (1999). What is river health? *Freshwater Biology, 41*(2), 197–209.

O'Donnell, E. (2018). *Legal Rights for Rivers: Competition, Collaboration and Water Governance. Earthscan Studies in Water Resource Management*. New York: Routledge.

O'Donnell, E., & Talbot-Jones, J. (2018). Creating legal rights for rivers: Lessons from Australia, New Zealand, and India. *Ecology and Society, 23*(1).

Oswald, A. (2012). *Dart*. London: Faber.

Palmer, T. (1994). *Lifelines. The Case for River Conservation*. Washington, DC: Island Press.

Park, G. (1995). *Ngā Uruora—The Groves of Life Ecology and History in a New Zealand Landscape*. Wellington: Victoria University Press.

Pessoa, F. (2002). *The Book of Disquiet*. London: Penguin.

Postel, S., & Richter, B. (2003). *Rivers for Life: Managing Water for People and Nature*. Washington, DC: Island Press.

Rapport, D. J., Costanza, R., & McMichael, A. J. (1998). Assessing ecosystem health. *Trends in Ecology & Evolution, 13*(10), 397–402.

Rockström, J., Steffen, W. L., Noone, K., Persson, Å., Chapin, F. S., III, Lambin, E., … Nykvist, B. (2009). Planetary boundaries: Exploring the safe operating space for humanity. *Ecology and Society, 14*(2), 32.

Rühs, N., & Jones, A. (2016). The implementation of earth jurisprudence through substantive constitutional rights of nature. *Sustainability, 8*(2), 174.

Ruru, J. (2018). Listening to Papatūānuku: A call to reform water law. *Journal of the Royal Society of New Zealand, 48*, 215–224.

Salmond, A. (2014). Tears of Rangi: Water, power, and people in New Zealand. *HAU: Journal of Ethnographic Theory, 4*(3), 285–309.

Salmond, A. (2017). *Tears of Rangi: Experiments Across Worlds*. Auckland: Auckland University Press.

Shiva, V. (2002). *Water Wars: Privatization, Pollution, and Profit*. Cambridge, MA: South End Press.

Singer, P. (1972). Famine, affluence, and morality. *Philosophy and Public Affairs, 1*(3), 229–243.

Smith, J. L. (2017). I, River?: New materialism, riparian non-human agency and the scale of democratic reform. *Asia Pacific Viewpoint, 58*(1), 99–111.

Steffen, W., Broadgate, W., Deutsch, L., Gaffney, O., & Ludwig, C. (2015). The trajectory of the Anthropocene: The great acceleration. *The Anthropocene Review, 2*(1), 81–98.

Stone, C. D. (1973). *Should Trees Have Standing? Toward Legal Rights for Natural Objects*. Los Altos, CA: William Kaufmann.

Strang, V. (2004). *The Meaning of Water*. Oxford: Berg.

Taylor, B., Van Wieren, G., & Zaleha, B. D. (2016). Lynn White Jr. and the greening-of-religion hypothesis. *Conservation Biology, 30*(5), 1000–1009.

Te Aho, L. (2010). Indigenous challenges to enhance freshwater governance and management in Aotearoa New Zealand—The Waikato river settlement. *The Journal of Water Law, 20*(5), 285–292.

Thomas, A. C. (2015). Indigenous more-than-humanisms: Relational ethics with the Hurunui River in Aotearoa New Zealand. *Social & Cultural Geography, 16*(8), 974–990.

Tousley, N. (2018). Tanya Harnett: The poetics & politics of scarred/sacred water. In J. Ellis (Ed.), *Water Rites: Reimagining Water in the West* (pp. 33–43). Calgary: Calgary University Press.

Vörösmarty, C. J., McIntyre, P. B., Gessner, M. O., Dudgeon, D., Prusevich, A., Green, P., ... Davies, P. M. (2010). Global threats to human water security and river biodiversity. *Nature, 467*(7315), 555–561.

Walter, R. C., & Merritts, D. J. (2008). Natural streams and the legacy of water-powered mills. *Science, 319*(5861), 299–304.

Wilder, M., & Ingram, H. (2018). Knowing equity when we see it: Water equity in contemporary global contexts. In K. Conca & E. Weinthal (Eds.), *The Oxford Handbook of Water Politics and Policy*. Oxford: Oxford University Press.

Wilson, E. O. (1984). *Biophilia*. Cambridge: Harvard University Press.

Wilson, G., & Lee, D. M. (2019). Rights of rivers enter the mainstream. *The Ecological Citizen, 2*(2), 183–187. Retrieved from https://www.ecologicalcitizen.net/pdfs/v02n2-13.pdf.

Wright, S. (2015). More-than-human, emergent belongings: A weak theory approach. *Progress in Human Geography, 39*(4), 391–411.

WWF (World Wildlife Fund). (2016). *Living Planet Report 2016. Risk and Resilience in a New Era*. Gland: WWF International.

Yates, J. S., Harris, L. M., & Wilson, N. J. (2017). Multiple ontologies of water: Politics, conflict and implications for governance. *Environment and Planning D: Society and Space, 35*(5), 797–815.

CHAPTER 2

The Socio-ecological River: Socio-economic, Cultural and Environmental Relations to River Systems

Abstract Profound changes in societal relations to rivers have taken place throughout human history. Intimate relations to waterways are an integral part of hunter-gatherer life. Early agricultural exploits and hydraulic civilizations developed along river systems. Water fuelled the Industrial Revolution, as humans asserted their authority over rivers through adoption of a command-and-control approach to river management. Moves towards an era of river repair in recent decades seek to address environmental damage. Underlying knowledge bases and perceptions of a healthy river reflect societal relations to rivers, shaping what it means to 'Find the Voice of the River' at any given place and time.

Keywords Agricultural revolution • Industrial revolution • Anthropocene • River repair • River management

2.1 INTRODUCTION: EMERGENCE OF THE ANTHROPOCENE

The earth is fast becoming an unfit home for its noblest inhabitant, and another era of equal human crime and human improvidence ... would reduce it to such a condition of impoverished productiveness, of shattered surface, of climatic excess, as to threaten the depravation, barbarism, and perhaps even extinction of the species. George Perkins Marsh (1864, p. 43)

© The Author(s) 2020
G. J. Brierley, *Finding the Voice of the River*,
https://doi.org/10.1007/978-3-030-27068-1_2

The present period of Earth history is increasingly referred to as the Anthropocene—a human-dominated world shaped by intensive and irreversible human impacts at the global scale. Steffen et al. (2011) characterize the Anthropocene in three steps. It began with industrialization, starting around 1750. The second step commenced following the Second World War in 1945. During this phase, referred to as "The Great Acceleration", population growth, urbanization, infrastructure development, escalation of consumption and the development of genetic engineering and synthetic biology went hand in hand with the collapse of ecosystems, exploitation of resources, biodiversity loss and global warming. This step, referred to by Wright (2004) as the collapse of time, continues into the twenty-first century. The third stage of the Anthropocene begins in the present as a period of growing awareness of human impact on the environment and initial endeavours to create global governance systems to manage humanity's relationship with the Earth System (Chapin et al. 2010). As yet, the effectiveness of these measures is seriously questionable. With societal relations to rivers at its core, this chapter explores how the Anthropocene came about.

Throughout history, access to water has played a key role in the development of human society. Although there is plenty of water to meet human needs at a global scale, it isn't always available in the quantity and quality that we require, in the right places, at the right time. As water is a renewable resource that is readily transported, diverted and stored, enormous efforts have been made to manage its supply and use. Efforts to make water available when and where we want it have exerted a profound impact upon river character and behaviour.

Hunter-gatherer societies have localized interactions with rivers, streams and waterholes, framed in relation to access to potable water. Given low population numbers and plentiful resources, impacts upon river health are incidental, characterized by secondary responses to burning and clearance of forests and use of grassland areas. As humans spread across the globe, in what Flannery (2010) refers to as our gypsy history, populations have grown considerably and societies have broken free of environmental constraints, asserting human authority over rivers in dramatic ways. Major efforts have been made to make drylands wetter through transfer and storage of water, and wetlands drier through drainage of swamps. Rivers supported the development of trade and commerce that underpinned the emergence of civilizations and cities. The development of industry not only transformed society, it also had a profound impact upon rivers, as a command and control ethos engineered rivers to meet societal needs. Through construction of dams, canals and channelized watercourses, rivers provided a source of energy, navigable channels and water supply and storage facilities. Increasing assertions of human authority over rivers further separated humans from nature. Growing populations and technological prowess accentuated human control over rivers. Collectively, these impacts left a profound legacy upon river systems, destroying many of the co-evolutionary bonds that lie at the heart of productive ecosystems (Best 2019; Macklin and Lewin 2019; Postel and Richter 2003). In recent decades, and in very different ways in different parts of the world, growing environmental consciousness prompted various calls and actions in moves towards an era of river repair (Brierley and Fryirs 2008).

Inevitably, the mix of biophysical, socio-economic, political and cultural considerations that shape societal relations to rivers varies markedly in space and time. These are issues of geographic specificity and historical contingency. Much depends on the type and scale of river, its position and environmental setting, its socio-economic and cultural context. Differing forms of co-evolution of societal development and river evolution took place in different areas over different timeframes, as rivers played different roles at different stages of human history (Chakraborty and Chakraborty 2018; see Box 2.1).

This chapter presents an overview of relations to rivers in hunter-gatherer, agrarian, industrial and post-industrial societies. Although material covered is generalized and selective, and is expressed in a linear

Box 2.1 Hey Gary, Where Are the People?
Following a talk on river diversity and variability at a workshop in Poland in 2013, the programme leader for the EU REFORM (REstoring rivers FOR effective catchment Management) Project, Tom Buijse, asked why none of my photographs included images of people. I wasn't consciously trying to convey a sense of the world as it was—historical relics that represent dreams of natural rivers (cf., Montgomery 2008), but all of the examples I had chosen excluded direct human impacts. Such selective and unintentional representations of river diversity are far removed from realities across much of our planet.

manner, it should not be construed as a linear historicist narrative that conveys transitions from 'traditional' 'hunter-gatherer' societies to 'industrial' societies through agricultural and mercantilist (or colonial) societies. As noted by Howitt and Suchet-Pearson (2006), a linear notion of space-time that flows in a unidirectional manner from past to future, conceptualizing reality as parts not wholes, fragments not continuity, is unthinkable to many indigenous peoples. All too often, breaking down time and space into measurable and manageable parts has been part of an armoury of practices that facilitate domination and control. Inevitably, multiple drivers and circumstances shape societal relations and activities in a contemporaneous manner, with trajectories that are recurrently disrupted and continuously remade.

2.2   Hunter-Gatherer Societies

Don't believe tree-huggers who claim that our ancestors lived in harmony with nature. Yuval Noah Harari (2014, p. 82)

The East African Rift can be considered as the original Silicon Valley—the place where tool-making hominids emerged around three million years ago (Dartnell 2018). Power and precision grips supported the use of stone tools and the emergence of a hunter-gatherer society. The first use of fire around half a million years ago supported more sedentary activities in a wider range of environments, helping to keep caves warm and big predators away. It supported cooking and smoking of meats, and the burning of undergrowth. The development of speech around 70,000 years ago marked a cognitive revolution, enhancing the capacity for groups of up to 150 people to collaborate in actions such as hunting, thereby playing a significant role in our shared ancestry (Harari 2014).

Climatically induced shifts in water availability and vegetation growth likely acted as a stimulus for human development. Greater availability of resources during wet periods may have supported population growth that instigated dispersal along the green corridors of valley floors. Fossil and DNA evidence suggest successive waves of human expansion and migration out of Africa (Fairbanks 2012). Limited capacity for food production and storage likely resulted in recurrent phases of dispersal and expansion—moving with tools in hand from one area to another, exploiting readily available resources along the way. Human movement was influenced by the changing seasons, the annual migrations of animals and the growth cycle of plants. An intimate knowledge of local resources was required to survive. Rivers and wetlands provided abundant freshwater and waterfowl, while coastal areas provided a ready source of seafood. Permanent fishing villages were established long before the Agricultural Revolution. Harari (2014) refers to pre-agricultural forager societies as 'the original affluent societies', characterized by a wholesome and varied diet, relatively short working hours and the rarity of infectious diseases. In a similar vein, Marr (2012) refers to early societal interactions within cooperative village life as a form of 'gentle anarchism'.

As relations to rivers were fundamental to human wellbeing, it is likely that societal knowledge was innately river-centric! Survival was dependent upon access to food, water and shelter. Instinct and intuition came to the fore. For hunter-gatherers, awareness of environments was innately and intimately experiential. Raymond Dart (1940, p. 178) contended that Australopithecus was … "a cave-dwelling, plains-frequenting, stream-searching, bird-nest-rifling and bone cracking ape who employed destructive implements in the chase and preparation of his carnivorous diet." Survival depended upon primal instincts and heightened sensitivity to surroundings in efforts to address concerns for hunger and safety. Rivers and wetlands doubtless served as primary sources of drinking water, for both humans and their hunting prey. Knowledge was distinctly local, with limited sense of what lay beyond a particular neighbourhood.

In seeking to explain why things happen, pagan animist framings attribute agency to non-human agents like rivers, the wind or the sun (White 1967). Belief in the supernatural provides an account of human existence that gives comfort in uncertain times, as ancestral spirits act as guardians that support societal wellbeing by controlling demons, providing guidance, reassurance and inspiration. With no separation between humans and nature in such enchanted visions, nature is animated by spiritual beings, considering fauna, flora, water, soil, rocks, landforms and rivers as living entities (Hoefle 2016). Almost every place, animal, plant and natural phenomenon is considered to have awareness, such that it communicates directly with feelings. All of nature, including the earth and sky, is alive, vital. In these ancestral relations, the spirits of the dead remain within forests, rivers and mountains, such that their presence is felt among the living. In such non-hierarchical ways of being, humans do not consider themselves to be dominant over other living things and non-human entities do not exist merely to serve human needs. Immaterial entities, the spirits of the dead, prevail alongside objects and living things as animated beings, with no barrier between humans and other beings.

It would be inaccurate and misleading to consider hunter-gatherer societies as ecological saints whose religious beliefs engender a reverence for nature (Budiansky 1995). The quest for survival reigns supreme over notions of sustainable living. Evidence from many fossil sites indicates that arrowheads became smaller over time, indicating hunting of progressively smaller beasts as the larger stock had already been eaten. As game died out, gathering of wild fruits and vegetables became increasingly important food sources. Altered fire regimes profoundly impacted landscapes and ecosystems. Across the world, human expansion brought about the near instantaneous demise of big game, including diprotodons and genyornis in Australia, mammoths and woolly rhinos in Europe and Asia, camels, mammoth, giant bison, giant sloth and horse across the Americas. Adaptation was the name of the game.

Various forms of environmental manipulation and management blur the distinction between foragers and farmers, hunter-gatherers and cultivators (Pascoe 2014). For example, large scale fishing facilities developed and applied by Aboriginal peoples in Australia exemplify river management systems that have continuously sustained healthy communities and cultures for thousands of years (e.g. Maclean et al. 2012; McNiven et al. 2015).

As co-evolutionary pathways were transformed, so were landscapes and ecosystems, in turn altering the ways in which human society interacted with them. In relative terms, small numbers of hunter-gatherers thinly spread over vast territories exerted limited and localized impacts on river systems in a galaxy of isolated human worlds (Harari 2014). Major rivers likely acted as barriers and borders to patterns of human movement.

By about 10,000 years ago, modern humans had migrated to all parts of the world. As human populations grew, the demand on local resources intensified and people were gradually less able to support themselves exclusively through hunting and gathering. This brought about a transition to intentional agricultural activities, cultivating wild plants and caring for wild animals, thereby changing societal relations to landscapes and ecosystems.

2.3   The Agricultural Revolution
and the Emergence of Hydraulic Civilizations

Although early domestication of plants and animals likely started during the Mesolithic and Neolithic period, human activities and interactions with the environment at this time were largely restricted to hunting, gathering and fishing. A shift to more ameliorable and predictable climatic conditions at the end of the Ice Age likely provided the stimulus for the transition from a hunter-gatherer society to a more sedentary lifestyle (Flannery 2010). As the development of agriculture supported prospects to stay in productive areas for longer periods, permanent settlements could be maintained so long as there was continuous availability of water, food and shelter. Streams and rivers played a critical role in these deliberations.

Available evidence indicates that the first steps towards sedentary life-styles were taken by the Natufians in recovering woodland areas of the Levant in the Middle East, where wild wheat, rye and barley flourished in transitional climatic conditions at the end of the last ice age. Dartnell (2018) describes this as a Garden of Eden for hunter-gatherers, who established villages of stone and wood, gathering wild cereals along with fruit and nuts from the woodland, and hunting gazelle. However, these conditions did not last. Climate perturbations induced by release of vast volumes of water to the Arctic Ocean during the Younger Dryas, around 13,000 years ago, triggered teleconnections whereby the relatively pleasant conditions in the Levant were disrupted by much colder and drier conditions that lasted for around 1000 years. This transitioned the homelands of Natufian society into arid treeless steppe conditions of thorny shrubs, diminishing the supplies of wild foods. Some Natufians abandoned their fledgling sedentary lifestyle and returned to migratory foraging. However, others were spurred to change their ways, developing agricultural exploits by purposefully planting seeds as the first step towards domestication.

The transition from gathering to gardening and from hunting to herding marked the emergence of agriculture in various parts of the world from around 11,000 years ago to 5600 years ago. The availability of differing species triggered non-synchronous developments, including wheat, barley and rye in the Fertile Crescent of the eastern Mediterranean, millet, soy bean and rice in China, corn, beans and squashes in Central America and Mexico, potatoes, manioc and tomatoes in the Andes-Amazon region of South America, sunflower in the woodlands of eastern North America and taro and banana in Papua New Guinea. Domestication of wild plants and animals through artificial selection, choosing the species that best suited particular conditions, marked a different type of evolution—the emergence of new species through intentional and direct human effort (Fairbanks 2012).

The oldest-dated villages were established in the hilly landscapes of southern Turkey around 11,000 years ago. Related practices then spread onto the plains of Mesopotamia, between the Tigris and Euphrates Rivers. The first human settlements were established in areas where fresh surface water was plentiful, adjacent to perennial rivers or springs. Settlement size was limited by the reliable and secure availability of freshwater. In areas of less reliable flow, techniques were developed to extract water from groundwater stores, supported by water transfer and storage facilities. In parts of the Middle East, Europe, India and China, permanent water wells have been used since around 8000 years ago.

Early agricultural exploits followed a similar dispersal pattern, starting in southern Turkey, then spreading to Mesopotamia around 7500–7300 years ago (Dartnell 2018; Marr 2012). Together with the Levant and the River Nile, this area makes up the Fertile Crescent. As this area saw the first farmers and the first large settlements, it is not surprising that it also gave birth to the first cities and the first empires.

Early approaches to floodplain farming utilized seasonal flow variation to supply water to agricultural land. Lifestyles were closely attuned to the flood pulse and the biological productivity of floodplains. Continuous cultivation practices along the Nile Valley in Egypt have been maintained for over 5000 years (Postel and Richter 2003). Formalized religions defined the relationship between natural and supernatural forces, conceiving regular flooding as a gift of the gods, designed to ensure the continuity of human life (Bowler 1992). Recurrent worship included singing of hymns to Hapi, the god of the Nile (Postel and Richter 2003). The river also provided a ready-made two-way transit system, with trade winds supporting up-river transport and trade, while river flow supported downstream movement (Dartnell 2018).

Excess food production and population growth, along with ready access to transport networks along navigable rivers, facilitated the development of trade, marketing and distribution of products. With wealth came measures of social and political organization—the emergence of a hierarchical society, including rulers, priests, merchants and landlords. The development and maintenance of civilizations built upon carefully organized and planned activities, with clearly defined roles and responsibilities and specialized skillsets. Technical and cultural advances included the use of pottery, metal, wheel, plough, clothing, shelter, speech and writing.

Hierarchical governance arrangements of hydraulic civilizations were structured around the development, use and maintenance of water supply facilities. These societies behaved like small nations, maintaining and defending their own territories or core areas (Diamond 2013). Enhanced communication underpinned moves to become stronger, smarter and better organized. Ancestor worship expressed reverence to the gods. In the world's first city at Eridu, raised pyramids called ziggurats were constructed around 7000 years ago overlooking the muddy, watery and sun-baked flatlands of Mesopotamia (Marr 2012). Greater stability engendered greater comfort. Babylonian and Persian architects developed large underground networks of irrigation tunnels to transport and store water through deserts, supporting life and even providing cooling facilities in hostile environments.

The emergence of settlements and civilizations along river systems marked a profound transition in human relations to the environment, shaping and manipulating the world rather than merely reacting to it. Plants and animals were turned from equal members of a spiritual round table into property (Harari 2014, p. 236). Economic concerns were increasingly considered alongside cultural relations. Human endeavours were conceptualized separate from an increasingly domesticated nature. Spiritually bifurcated worldviews were characterized by subordination and submission to all-powerful gods, as gnostic belief systems asserted that absolute religious truth can be known, revealed through enlightened humans who act as representatives of God, acting on his behalf to make use of material products created in his vision (Hoefle 2016). Spirits came to be divided into rival groups representing good and evil. More intensive and unequal use of natural and human resources brought about the accumulation of wealth and greater inequality between social classes.

Prior to the first agricultural revolution, human impacts upon the environment were piecemeal, fragmented and localized. Subsequently, land use and water management practices brought about systematic and increasingly profound adjustments to landscapes and ecosystems, in terms of both scale and intensity. Use of fire, stone tools and metalwork cleared forests, ploughed lands and drained swamps.

With rivers at their core, hydraulic civilizations developed independently at strategic locations in various parts of the world (Wittfogel 1955). Civilizations in Sumer (southern Mesopotamia, modern-day Iraq) and Egypt emerged around 5000 years ago. By 3000 years ago, civilizations had arisen in at least five more places: The Mediterranean, India, China,

Mexico and Peru. Complex systems of drains and dykes, along with dams and levees, supported irrigated agriculture. With the emergence of hydraulic civilizations, rivers were increasingly seen as resources to be managed and manipulated to support human needs.

In China, a legendary leader Da Yu (c. 2123–c. 2025 BC), the founder of the Xia dynasty, became known as Yu the Great because of his coordination of efforts to tame the floods of the Yellow River (Marr 2012). Building on the failed efforts of his father, Yu used thousands of kilometres of canals as well as earth dams and dikes to beat the endless floodwaters of the river, dividing the flow and dissipating its energy among multiple channels. Elsewhere in China, flow and sediment management along the Min River at Dujiangyan in Sichuan Province has maintained continuous operation of a large irrigation scheme that has fed the Chengdu Plain for over 2400 years. Dramatic land use change, drainage of swamps and river diversion schemes transformed middle reaches of the Yangtze River. Increasingly sophisticated water engineering projects included construction of the 1800 km long Grand Canal between Hangzhou and Beijing around 2500 years ago.

Careful orchestration of activities was required to supply food and water to increasingly urbanized societies, storing sufficient food to survive through periods of flood and drought or in response to threats from other societies. Management of water quantity and quality had a significant impact upon human health and wellbeing. Where water resources, infrastructure or sanitation systems were insufficient, diseases spread and people fell sick or died prematurely. Typically, dispersing and diluting raw sewage in rivers provided a crude form of sewage disposal facility. Temples at Nippur and Eshnunna in Babylonia first used clay sewer pipes around 6000 years ago. The first sanitation systems were built in prehistoric Iran, near the city of Zabol. Reuse of domestic wastewater for irrigation purposes was integrated into water management systems in places such as Mesopotamia and the Indus Valley. Use of aqueducts and indoor plumbing by the Minoan civilization at Knossos in Crete not only provided sanitation facilities through sewer systems, but also bathing facilities to support therapeutic, recreational and aesthetic interests. At Mohenjo-Daro along the Indus River in modern-day Pakistan, the Harappan civilization that flourished there since around 5300 years ago developed rainwater harvesting techniques and devices to lift water to ground level to support sophisticated irrigation structures. Notable advances in urban water management included public and private baths, covered drains on the main streets and a sophisticated sewage disposal system comprised of brick-lain underground drains that connected individual houses with private toilets to wider public drains.

Advances in bridge construction techniques altered human relations to rivers. Simple timber pile and crude woodwork structures have been used to cross smaller rivers since around 3600 years ago. Beam or girder structures made from wood and stone crossed the Euphrates River in Mesopotamia at least 2800 years ago. These early engineering achievements supported the growth and connectivity of cities, increasing the scope of trade and business networks.

Eventually, most hydraulic civilizations collapsed (Diamond 2005). For example, various cities that made up the Sumerian civilization in Mesopotamia, such as Babylon, Uruk and Ur were abandoned between 6500 and 3900 years ago. Initially, food security problems likely reflected famine as a result of drought conditions. The limited capacity to store or transport food meant that it was difficult to cope with anything other than a minor crisis. In turn, crop failure, disease and overpopulation brought about social disarray. Over time, upstream deforestation and erosion induced sedimentation and flooding problems in downstream areas. Overgrazing and land degradation reduced soil fertility. Salinization was the final straw. Desertification—the emergence of human-induced deserts—reflected limited social and organizational capacity to plan proactively and respond appropriately to impending crises, failing to respond to progressive environmental degradation. An inability to sustain the supply of water and food engendered collapse.

By around 2000 years ago, the Roman Empire emerged as the world's first recognizably modern culture (Flannery 2010). The Romans maintained and enhanced intellectual traditions established by the Greeks, with the rise of Christianity paving the way for a new spiritual and intellectual era (Bowler 1992). At this time, Rome comprised a sophisticated multicultural society, with outstanding infrastructure and water engineering structures such as aqueducts and bathing facilities. Concrete and mortar supported the construction of much larger dam structures than previously. Hydropower generation using waterwheels powered water-raising mechanisms that fed networks of aqueducts and pipes. Early variants of channelization programmes, tunnel development and hydropower schemes supported the use of rivers for navigation and land development/reclamation schemes. The construction of Pons Sublicius enabled the Romans to cross the Tiber River in Rome around 2800 years ago. Enhancing various techniques developed in Greece, the Romans built many truss and arch bridges, the latter being much lighter than previous structures and able to hold loads that were twice as heavy as the bridge itself. A major sewage system, the Cloaca Maxima, was first constructed around 2600 years ago, draining local marshes and transferring effluent from Rome to the river. Early versions of cesspools were forerunners of modern-day septic tanks, using settling properties of materials to return partially cleaned liquids into nearby waterbodies. Collected sludge was either used as fertilizer or simply buried. Multiple aqueducts which supplied water to Rome about 1900 years ago supplied many public baths, public fountains, imperial palaces and private houses before being channelled into the sewer network under the streets of the city. Rome collapsed around 1500 years ago when its food supply was cut and aqueducts were severed.

The agricultural revolution, adoption of sedentary lifestyles and permanent settlements and the emergence of hydraulic civilizations reflected profound transformations in societal relationships to land and water. Changes to knowledge and belief systems accompanied this transformation. Although the scale of activities remained relatively localized, environmental impacts were especially pronounced along valley floors, shaped by concerns for the provision of resources to meet human needs. While rivers previously acted as boundaries and borders to human activities and movement, the expansion of commerce increased their use as conduits for transfer and trade. Strategic use of rivers supported defence mechanisms for cities and their emerging civilizations. Conversely, valley floors provided important routes for migration and/or invasion. Growing populations and capacity to respond to environmental changes started the boom and bust cycle of socio-economic circumstances. In some instances, climate change triggered moves towards more intensive irrigation schemes. Elsewhere, if there was insufficient capacity to respond to prevailing pressures, collapse ensued.

Various scholars in ancient China and Greece wrote about degraded environmental conditions as a result of human activities several thousand years ago. For example, Hippocrates (c. 460–c. 370 BC) recognized distinct linkages between environmental conditions and human health in his text entitled *Airs, Waters, Places*. Subsequent scientific and industrial revolutions brought about further expansion of human activities and impacts, characterized by more intensive, broader-scale changes to landscapes and ecosystems.

2.4   CHANGES IN SOCIETAL RELATIONS TO RIVERS IN RESPONSE TO THE SCIENTIFIC AND INDUSTRIAL REVOLUTIONS

The analysis of Nature into its individual parts, the grouping of the different natural processes and objects in definite classes, the study of the internal anatomy of organized bodies in their manifold forms—these were the fundamental conditions of the gigantic strides in our knowledge of Nature that have been made during the last 400 years. But this method of work has also left us as legacy the habit of observing natural objects and processes in isolation, apart from their connection with the vast whole; of observing them in repose, not in motion; as constraints, not as essentially variables; in their death, not in their life.
Frederick Engels (1910, chapter 2, p. 1)

Prior to A.D. 1500, organically framed worldviews of a 'living' universe emphasized the interdependence of spiritual and material phenomena and the subordination of the individual interests to those of the community (Capra 1982). Inter-twined assertions of reason and faith focussed attention upon the meaning and significance of things (i.e. their purpose), rather than endeavours to predict or control them. Subsequently, the scientific revolution conceptualized the world as a machine and this has become the dominant metaphor through to contemporary times.

Building upon scientific developments in previous centuries, the onset of the Industrial Revolution around 1750 brought about the most significant change in societal relations to landscapes and ecosystems since the domestication of animals and plants, liberating humankind from its dependence on the surrounding ecosystem (Harari 2014). Scientific and technological developments supported rapid growth in commerce and trade. Population growth and urbanization were accompanied by major changes to social structure and lifestyles—jobs, homes, ways of living. Approaches to knowing and managing rivers were transformed as applications of technical and engineering prowess sought to control and exploit rivers to meet human needs.

Greek and mediaeval knowledge frameworks sought ways to attain the supreme truth, whether of a theological or ontological order. Mediaeval beliefs conceptualized nature as hidden and unfathomable, fashioned by supernatural explanations. Ways of knowing emphasized the duality of man and nature, wherein human activities were guided by the Creator, applying religious belief structures that envisaged 'Man' as being created in God's image. Notably, in the Christian faith, human inheritance of the Earth validated prospects to exploit nature in a mood of indifference to the feelings of natural objects (Bowler 1992). Following the Scientific Revolution, belief structures based on metaphysics, mysticism, magic and alchemy were progressively replaced by rational scientific practices. While science is typically conveyed as an objective quest for truth, always willing to abandon a given idea for a better one, religion is often conceived as an unthinking allegiance to dogma and obedience to authority, holding its truths to be eternal and immutable. White (1967) contended that Western scientific and religious ideas, working in concert, precipitated the ecological crisis.

As nature became the object of scientific investigations concerned with physical and material phenomena, metaphysical questions regarding philosophy and theology were increasingly set aside. Notions of coexistence, reciprocity and mutual interdependence became marginalized in the separation of human society from nature, mind from matter, the material-physical phenomenon from the spiritual-metaphysical and organisms from inorganic-non-vital matter. In this secular materialist vision, humans came to see themselves as superior to the plant and animal kingdom, but subordinate to the controls of gods, saints and angels (Hoefle 2016). Through such framings, acting in the image of God was considered to give humanity supreme control over the natural realm.

Several scholars made quite alarming references to nature at this time. In his book *The Sacred Theory of the Earth*, Thomas Burnet (c. 1635–c. 1715) built upon biblical interpretations of the earth's history to express beauty in terms of orderliness and regularity, casting aspersions of a 'dirty little planet' that was overgrown by ugly and dangerous mountains. Theological tensions highlighted conflicts between those who believed that the world was designed by God and those who saw it as a ruined planet fit for sinners (Bowler 1992, p. 112). Prior to the Romantic Movement, landscapes and the environment were far from admired and revered in the Arts, as proponents of popular culture bemoaned the woes and evils of mountains, forests and wetlands. For example, John Donne (1572–1631) described mountains as 'warts', while George Leclerc (Comte de Buffon, 1707–1788) described wetlands as 'putrid and stagnating waters' and forests as 'a disordered mass of gross herbage' (quoted in Budiansky 1995).

Seen through a mechanistic, reductionist and rationalist lens, nature came to be conceptualized as a machine in which specialized and interdependent parts are governed by universal and immutable laws which can be described by the human observer using physical-mathematical language. Nature was viewed as profane, something that can be tinkered with and radically rearranged without destroying its organic integrity, because implicitly it was not considered to have any (Bowler 1992). Merchant (1980, p. 169) draws attention to early scientific aims to "torture nature's secrets from her".

Technical-scientific evolution fed the economic-productive model of the bourgeois, fomenting the capitalist system. Building upon the Protestant work ethic, success in business became a visible symbol of spiritual worth, because "God intended us to use his energy and initiative to exploit the gifts of His creation, (wherein) the search for practical knowledge was in itself a religious activity" (Bowler 1992, p. 88). Increasingly, European society came to be governed by an elite that drew its resources from trade rather than the land. An emerging middle class of industrialists and businessmen came to replace a landed class of nobility and gentry. Significant increases in personal wealth lead to a consumption revolution. For the first time in history, there was a simultaneous increase in both population, which started to grow rapidly in the nineteenth century, and per capita income.

This new social order supported and embedded new forms of commercial and industrial power that allowed Europe to dominate the globe. The quest for profit was a powerful motivating force in the colonization of new lands. Little consideration was given to opening up new lands in ways that managed water, timber and soil wisely as sustainable resources to support the wealth of the nation, providing the greatest good to the greatest number, relative to plundering them in the name of a quick profit by a few selfish plutocrats (Budiansky 1995). The new materialism was justified and vindicated by a sense of divine approval in efforts to take advantage of the resources created by God solely for that purpose (Bowler 1992).

Where nature had once been seen as an organic whole, a source of mysterious constructive powers, the material world was portrayed as a passive, mechanical system that had no moral dimension. An ideology of utility came to the fore, manipulating elements of the natural world at will without worrying about broader implications. Nature came be seen as a vast, carefully structured machine based on the laws of economics. The fact that so many species were of use to human endeavours suggested that the Creator had made limitless resources available, to be manipulated for societal benefit. The balance of nature was stable by God's decree—yet it was flexible enough to allow human activities to multiply the numbers of those species of value to society and to destroy those which are pests. A materialistic view of nature was very much in keeping with the demands of the new industrial society. As science supported the expansionist interests of commercial and/or military concerns, it became an indispensable tool for world domination in an increasingly industrialized world, exuding a form of moral authority (Bowler 1992). Knowledge became increasingly specialized and discipline-bound, developed and applied by experts using increasingly elitist language. Geology helped mining and resource extraction. Botany and zoology supported farmers, fishermen and hunters. Selective breeding of crops and animals and pest control measures supported industrial-scale monocultural agriculture. The development of the steamship, refrigerated shipping and the telegraph supported trade and communication networks. Quinine provided a major boost for work in the tropics. More efficient water wheels and steam power transformed the availability and use of energy. Transportation became faster and cheaper. New forms of bridge construction used iron and steel as part of truss structures, and later as part of cantilever and suspension bridges. The production of cement, and the rediscovery of concrete after a hiatus of around 1300 years, provided enormous support for the building trades, facilitating the construction of houses, factories, industrial buildings, infrastructure, dams, tunnels and sewerage systems.

Industrialization brought about profound transformations in societal relations to rivers. Rivers were used as tools for transport, to fuel industry, and as gutters, drains and sewers to wash away untreated waste products. Rivers such as the Mersey in northwest England and the Rhine (Ruhr) in Germany became the backbone of industrial society. Many river networks were transformed in rapidly expanding urban areas. Often cities turned their backs to the river.

Despite the technical prowess of earlier civilizations, most cities did not have a functioning sewer system before the Industrial era. Rather, they relied on rain showers to wash away the sewage from the streets to nearby streams and rivers. Notable expansion in waterworks and pumping systems was undertaken in the seventeenth and eighteenth centuries. In the late nineteenth and twentieth centuries, municipal sanitation programmes constructed extensive sewer systems to help control outbreaks of disease. During the Industrial Revolution, the River Thames in London was identified as being thick and black due to sewage, such that the river "smells like death" (Ackroyd 2007). A map generated by John Snow in 1854 showed that clusters of cholera were related to water supply facilities. Following the *Great Stink* of 1858, an extensive network of underground sewers and pumping stations was constructed as part of a wastewater treatment system. However, it was only in the half-century around 1900 that public health interventions drastically reduced the incidence of waterborne diseases among the urban population.

By the beginning of the twentieth century, European nations had claimed most of the habitable world as colonies. Extensive and intensive changes to landscapes and ecosystems took place virtually instantaneously. In his book entitled "Guns, Germs and Steel", Diamond (1997) outlines how guns asserted power relations, disease decimated indigenous populations and mechanized technologies enabled rapid and widespread forest clearance and land use change. Direct efforts to tame the land were accompanied by the introduction and spread of exotic species of plants and animals.

By the end of the nineteenth century, the Darwinian revolution brought about a different form of stimulus that further accelerated the exploitative underpinnings of a development ethos. In dismantling notions of the mystical unity of nature as a balanced and stable entity fashioned by a benevolent Creator, human interference with the natural world was reconceptualized as part of a constant struggle based upon principles of natural selection (Bowler 1992). Assertions of the survival of the fittest perceived nature as rewarding those who were able and energetic, and punish those who could not keep up with the race towards higher things. This created a sense that human beings must respond to the challenge of the environment in the same way that biological evolutionism in nature reflected the constant struggle of organisms. In this age of progress through struggle, free enterprise was seen as the key to social progress. Industrialists exploited their workers, and western nations exploited the rest of the world, as power became increasingly concentrated in the hands of the captains of industry. Domination of human-over-nature was accompanied by domination of human-over-human. Imperialist and globalizing traits resulted in profound social exclusion, poverty, inequality, conflict/war and environmental degradation.

Carefully crafted institutional arrangements underpinned approaches to knowledge generation and use. Beattie and Morgan (2017) conceptualize these relations in the British Empire as "A Rivered Eden", wherein functional, utilitarian knowledges underpinned the imposition of a command and control approach to river management. Engineering applications were developed, trialled, implemented and transferred across the colonies, with little regard for local diversity and variability. These externalized knowledge structures have asserted a significant institutional legacy (Beattie and Morgan 2017). The imprint of the government-sponsored (public agency) work that underpinned the opening-up of colonial landscapes remains to this day. Sadly, imposed scientific knowledges were seldom related in meaningful ways to traditional knowledges of indigenous peoples, gaining insights from long-learned ways of living with the land. Even more tragically, colonial interactions fractured indigenous connections to landscapes and history.

A command and control approach transformed rivers in both the Old and New Worlds. Major water engineering projects significantly reshaped societies, economies and ecosystems. Government engineers were tasked to create irrigation, flood control, hydropower and water supply facilities and navigable channels (Hüttl et al. 2016). Major advances in technological prowess greatly increased the scale and scope of efforts to tame, train, regulate and control rivers through various forms of channelization, canalization and dam construction.

Channelization enhances the capacity and flow of the river, supporting navigability by large vessels with deep draughts. It also increases human access to a greater proportion of the floodplain, initially for agriculture, but subsequently for a range of urban and industrial uses. In such endeavours, a uniform, smooth, straight, larger capacity channel is constructed to replace the former more sinuous, often multi-channelled course. Typically, the channel bed and banks are lined with concrete to enhance stability and minimize erosion. Artificial levees, sometimes called dykes or stopbanks, train the channel, abandoning floodplain surfaces with their cut-off channels, palaeochannels and wetlands.

Canalization measures regulate channel flow with a particular grade (slope) such that adequate and continuous flow depth is maintained. Water is impounded behind weirs to regulate both flow and slope, while dykes close-off subsidiary low-water channels. Often, deflection structures direct water into a central channel at low flow stage. Impediments to the modified channel, such as rapids at local obstructions, are removed or an alternative course is constructed. Locks alongside weirs, or in side channels, enable the passage of vessels through a succession of fairly level reaches that rise in steps upstream. Rivers such as the Danube, Rhine,

Mississippi, Mekong and Yangtze became major highways for inland navigation and transport, aiding movement of people and especially goods for trade. In addition, extensive canal networks that support irrigation programmes have transformed many low relief landscapes.

Large dam projects generate hydroelectric power, support large scale irrigation schemes, aid navigability and are an important part of flood protection plans. Over 60,000 large dams (greater than 15 m from foundation to crest) have been constructed across the world. Since the late twelfth century, dams have held back water to support the development of agriculture on previously low-lying marshlands in parts of northern Europe. Many of these developments became crossing points of rivers, supporting the growth of towns and cities such as Amsterdam and Rotterdam. Masonry dam design in the latter half of the nineteenth century supported the construction of much larger dams. Construction of the Aswan Low Dam along the Nile River in Egypt in 1902 followed earlier and smaller developments in India, Canada and Australia. Building upon experiences gained in major engineering schemes along rivers such as the Rhine, the Danube, the Mississippi and the St Lawrence Seaway, by the mid-twentieth century virtually all large rivers had been subjected to a significant degree of human control (e.g. Petts 1984). The Hoover Dam in Colorado was built between 1931 and 1936 during the Great Depression to control floods, provide irrigation water and produce hydroelectric power. Reisner (1986) presents a compelling exposé of the bitter rivalry between two government giants, the Bureau of Reclamation and the U.S. Army Corps of Engineers, to dam and divert rivers in efforts to transform the American West to create an Eden. The Tennessee Valley Authority in the United States, the Snowy Mountains scheme in Australia and the James Bay Project in Canada systematically modified river systems in the interests of cheap and renewable hydropower and provision of irrigation water to enhance agricultural production. An unprecedented boom in dam and reservoir construction following the Second World War reflected the growing demand for irrigation, water supply and provision of hydroelectric power in response to the needs of growing populations and economic development.

Nationalistic endeavours through state-controlled programmes also brought about major transformations to rivers in countries such as China, India, Brazil, South Africa and the former Soviet Union. The scale of some of these endeavours is extraordinary. For example, the Three Gorges Dam along the Yangtze River in China, which commenced operation in 2012, raised water level 110 m above the downstream channel, with the reservoir extending 660 km upstream. The Itaipu Dam on the Paraná River on the border between Brazil and Paraguay is the highest energy-generating dam in the world. Completed in 1984, the dam is 196 m high and its reservoir covers an area of 1350 km^2. Extreme variants of a command and control ethos were expressed during Chairman Mao's War on Nature in China and Stalin's Plan for the Transformation of Nature in the Soviet Union.

All too often, megaprojects that are framed and marketed as programmes to meet societal needs serve the economic, utilitarian and material interests of opportunistic developers, engineers and the construction industry, rather than local residents, compromising societal dependency upon the benefits from healthy ecosystems. Typically, measures of sociocultural and environmental harm are unrecognized or unvalued and are omitted from benefit-cost analyses. Although decisions made today are fully cognizant of impacts and legacy effects upon aquatic ecosystems and displaced communities, and it is widely recognized that long-term economic benefits are questionable, several river basins are experiencing a new wave of damming and water-transfer schemes: the Amazon, the Congo, the Mekong, the São Francisco, the Magdalena, the south-north programme in China and several initiatives in India.

Imposition of a technocentric mindset sets rivers on a trajectory from which there is no going back. Path dependencies are difficult, if not impossible, to revoke. Such actions discount the future, emphasizing short-term benefits to the neglect or denial of long-term costs and consequences (Flannery 2010). The operational life expectancy and maintenance of facilities are significantly over-looked and under-estimated. All structures have a design life. Making existing dams bigger and stronger, or decommissioning and removing them, are extremely expensive options. Reinforcement, maintenance and expansion of dams, dykes and levees is a year-by-year imperative along many rivers. The Mississippi River in the United States expressed its voice during Hurricane Katrina in 2005. Tragic circumstances ensued as the river broke through levee banks near New Orleans. Subsequently, levee engineering teams returned the channel to its allocated place. Sadly, such disasters provide all too frequent reminders of the realities and legacies that have been imposed upon rivers (see Box 2.2).

Box 2.2 Waiting for Nature to Fight Back

Engineering structures have silenced the voice of countless rivers. In 2017, the Edgecumbe River in New Zealand temporarily escaped its stopbanks following a storm, only to be put back in its place. Disturbance events are common in the Shaky Isles, where many rivers are restrained within stopbanks. Examples in the Ruamahanga Catchment in Martinborough are especially alarming, with some channels perched several metres above the adjacent floodplain. Many of these channels flow along fault lines. Disasters are inevitable. It's not a matter of if, but when. In April 2019, the Waiho River at Franz Josef Glacier on the South Island took out the bridge and breached one of the stopbanks. The West Coast Regional Council has already agreed to put them both back, requesting national support as sufficient funding is not available in the regional tax base.

Ongoing commitments to restrain rivers are really expensive. Once such actions are taken, path dependencies are set, shaping societal expectations, such that it is very difficult to revoke such measures. Nature fights back. Eventually, it wins. When will we learn to 'Stop Doing Stupid Things'? Eventually, it is likely that the (re) insurance industry will sort such concerns, as making a profit is a powerful motive for the decisions and actions that they take.

The emergence of a human-dominated world of the Anthropocene reflected the pursuit of growth in the interests of progress. Such developments came at a price. Despite considerable knowledge of the damage done, expansion continued until intensive impacts had been achieved at a global scale. Economic interests suppressed concerns for cultural and environmental values. Water wars are an ongoing reality (e.g. Shiva 2002). Intensification of measures is now the name of the game. However, such moves have not stopped the push to 'develop' the last remnants. In barbarous acts of bastardry, exploitative interests seek to cut remaining areas of old growth forest, even if supposedly protected in conservation areas. Any fair-sized river that is perceived to have excess flow continues to be subject to pressure for 'development' (Hüttl et al. 2016). As cumulative impacts build inexorably, future options are increasingly limited. It is hard to underplay the extent of human transformation of riverscapes. A recent synthesis of direct human impacts in Britain revealed an average of one anthropogenic structure for every 1.3 km of river course (Jones et al. 2019). Collectively, these various forms of human endeavour have created a layer upon layer effect of human manipulation of rivers. The cumulative record of what has gone before constrains the range of future options. By the 1960s, a dawning environmental movement created a groundswell of support for environmental action, drawing attention to Only One Earth and the existence of Limits to Growth.

2.5   An Emerging Era of River Repair

Greed and selfishness can be thought of as essential ingredients of command-and-control systems: the Fat Controller is not merely a fiction from a children's story; he's fat because he controls. Tim Flannery (2010, pp. 217–218)

The ravages of the Second World War and economic pressures of previous decades left the world in quite a mess in the 1950s. Many riverscapes were in a shocking condition. Industrial landscapes were squalid environments, as pollutants, toxic waste and other factors presented countless problems. Sewage treatment facilities in most urban landscapes were not up to the job. Agricultural landscapes had minimal riparian vegetation cover. There was intensive competition for disconnected floodplain areas, while flows along adjacent regulated rivers were over-allocated. These working landscapes were far removed from carefully manicured, domesticated settings. Rather, they were depressing and unhealthy places. This was an age of disconnect and disrespect, for the environment and each other. Recognizing the damage done, time was ripe for an era of river repair. Although an emerging environmental movement facilitated such a transition in many parts of the world, other areas remain in an awful state.

The emergence of an era of river repair marked recognition of the imperative to better balance societal demands with the needs of rivers themselves, moving beyond a singular focus upon economic concerns through incorporation of environmental values as part of the mix. In her groundbreaking book, *Silent Spring*, Rachel Carson (1962) drew attention to the rapid dispersal of toxins in the atmosphere, on land and through river systems, as industrial practices and excess pesticide and fertilizer use set society on a cataclysmic course. Since the 1960s, a growing social movement has increasingly challenged the philosophy of environmental control and exploitation, questioning the right of humankind to tear the earth apart for its own material benefit. Alternative frames of reference and ways of thinking emerged as parts of a growing sustainability agenda. The Club of Rome (initiated in 1968) and the Brundtland Commission (initiated in 1983) called for socio-economic and political shifts to address concerns for environmental degradation associated with pollution, over-exhaustion of resources, population growth, food and water availability and loss of habitat, biodiversity and ecosystem functionality/integrity. Significant conservation and rehabilitation programmes have ensued.

The era of river repair is largely a product of an emerging environmental movement and associated green politics. Back in the 1960s and the 1970s, various visionaries promoted the development of social and environmental activist groups such as Greenpeace, Friends of the Earth and Amnesty International. Concerns for social and environmental justice lie at the heart of major campaigns to address issues of inequity. Given the sensitivity and vulnerability of waterways, concerns for the condition of aquatic ecosystems emerged as focal points for many environmental campaigns in the 1970s and 1980s.

Perhaps inevitably, under the circumstances, environmental repair was conceived as a 'one step at a time' process, in efforts not to be overwhelmed by the enormity of the task at hand. Discipline-bound practices provided the available frame of reference, supporting the management ethos of 'one problem at a time' (see Chap. 3). Concerns for air and water quality were the most visible and tangible problems of the day. As a starting point, legal and policy framings tackled concerns for industrial pollution and sewage treatment. Simply stopping dumping waste products into rivers brought about profound transformations. Although this did not happen overnight, progressive improvements were quickly evident. Environmental concerns became a focal point of attention, rather than a marginal issue (see Box 2.3).

Box 2.3 Transforming the Thames River from a Foul-Smelling Sewer ...

Although enhanced sewage treatment facilities constructed during the latter decades of the nineteenth century addressed immediate health issues along the Thames River in London induced by pollution and water quality problems, by the 1950s the river was a vast, foul-smelling, lifeless drain—a biologically dead open sewer (Ackroyd 2007). Further improvements to the sewage system, along with tighter controls upon pesticide and fertilizer use and management of toxic metal pollutants in the 1970s and 1980s, helped to regenerate the river such that it was able to breathe again. From virtually no fish in the 1950s, there are now 125 species of fish in the Thames. Similar approaches to management of industrial pollutants brought about marked improvements in river health in the Mersey Basin in north-west England and the Rhine River in the Ruhr region of Germany— the heartlands of the industrial era.

Since the 1980s, greater environmental consciousness has underpinned an era of increasingly green politics, with notable shifts in the policy arena and the corporate sector. Green business has become an entity in its own right, striving to meet the demands of a growing green marketplace. Environmental agendas frame State of Environment reporting procedures and monitoring programmes, forming an important part of consents processes that seek to minimize environmental harm. Over time, such programmes have moved away from top-down government-centred initiatives to incorporate a suite of participatory practices as part of a growing movement for societal engagement in the process of environmental repair. Environmental issues are increasingly incorporated within policy developments and corporate sustainability initiatives as part of corporate responsibility programmes.

Improved river health brought about a transformation in societal relations to rivers, reframing expectations of river condition (i.e. what was considered acceptable). Society had been living with unhealthy, often dead or sterile rivers for decades or centuries. Once society became aware of the benefits of cleaner water and the initial vestiges of a returning aquatic life, more people spent more time along river courses, raising expectations of enhanced conditions and the requirements of a duty of care to achieve and maintain this. There was no going back—the baseline of societal expectations had shifted in a positive way. Subsequently, the designation and implementation of environmental flows and the efforts of a burgeoning river restoration movement engendered much healthier rivers relative to decades past.

Much has already been achieved in efforts to improve river health across large parts of our planet. Legislative changes and policy framings increasingly support and underpin environmental programmes. Responses and approaches have been distinctly situated, in space and time (Brierley and Fryirs 2008). Much depends upon who sets the visions, designs the plans, implements the work, conducts monitoring programmes and how these processes are governed (Chap. 3). Some continue to perceive human dominance over nature as manifest destiny, perpetuating the myth of human dominion over nature by imposing anthropocentric values in managing and restoring rivers to a particular goal or endpoint. Many of these applications have an economic focus. Others seek to live in harmony with nature, seeing humans as part of nature, applying cooperative framings that revive and rekindle traditional knowledges that are part of our cultural heritage through approaches to biocultural restoration (Capra 1982, p. 412). Although laudable improvements in river condition have been achieved in local terms, prevailing practices are not working at the scale or the rate that is required. Ultimately, problems must be tackled at the scale of formative processes and drivers, remembering that activities on the land exert a primary influence upon river condition (Chap. 4).

2.6   Concluding Comment: Changing Perceptions of Healthy Rivers

The universe is no longer seen as a machine, made up of a multitude of separate objects, but appears as a harmonious indivisible whole; a network of dynamic relationships that include the human observer and his or her consciousness in an essential way. Fritjof Capra (1982, p. 47)

Over millennia, small, simple cultures have evolved and coalesced to form bigger and more complex civilizations, as merchants, conquerors and prophets created mega-cultures with distinctive economic, political and religious traits (Harari 2014). The growth of empires resulted in the spread of ideas, institutions, customs and norms, creating a form of cultural inheritance. Societies that avoided collapse came to make up today's globally interconnected world (Friedman 2006).

Table 2.1 presents an overview of societal relations to rivers, relating knowledge bases that underpin uses of rivers to impacts upon them. These relations determine what is considered to be a healthy river. In hunter-gatherer societies, functional access to water is a key part of day to day relations, living with variability and moving as required. Lifestyles are inherently local, with many rivers acting as barriers, boundaries and borders that constrain human movement. Wisdom gained from experience

Table 2.1 An overview of societal relations to rivers

Theme	Hunter-gatherer	Agricultural revolution	Scientific and Industrial revolution	Electronic and Information age
Approaches to knowing and relating to rivers	Informal and experiential, applying a live and learn, trial and error approach. Enchanted, holistic and animist visions see rivers and inanimate elements of the natural world as living entities, imbuing them as spirits and guardians	Principles of mysticism, metaphysics, alchemy and magic underpin relationships to rivers. Religious belief systems convey assertions of good and evil	Separation of mind and matter, humans and nature. Formal scientific and technical knowledge manipulate rivers to serve human needs, striving to make rivers stable and predictable entities. Rivers are conceptualized as machines. Discipline-bound science and assertions of faith impose human dominion and authority over rivers	A quest for integrative and holistic framings reconceptualizes societal relations to rivers. Catchment-specific understandings recognize and seek to work with uncertainty and emergence, framing scientific insights alongside local understandings in a multiple knowledges ethos
Economic, environmental and socio-cultural values that underpin uses of the river	Access to potable water is a primary driver. Reverence for water is expressed through animist connections. Adaptation is the norm, with success or failure shaped by prevailing conditions (e.g. floods and droughts)	Water is conceived as a resource to support agricultural developments. Rivers are modified to meet human needs. Agricultural surpluses underpin trade and commerce, supporting population growth	Command and control engineering practices assert human authority over rivers to serve economic interests, pushing aside socio-cultural and environmental values	Awareness of environmental damage prompts an era of river repair. Two key options are available. First, economically framed practices apply an ecosystem services lens. Second, a socio-cultural lens emphasizes concern for equity and social and environmental justice, applying a more-than-human ethos to support the rights of the river

(continued)

Table 2.1 (continued)

Theme	Hunter-gatherer	Agricultural revolution	Scientific and Industrial revolution	Electronic and Information age
Impacts upon the river	Localized, indirect and negligible impacts because of low population numbers and low intensity activities	Indirect impacts upon runoff and sedimentation regimes due to deforestation, use of grasslands and drainage of swamps. Increasing intensity of use of floodplain areas. Many hydraulic civilizations collapsed due to environmental and organizational problems	Alongside extensive indirect impacts, profound and intensive direct impacts occur over large areas. Water management megaprojects transform rivers. Dams and flow regulation structures store and transfer water to support irrigated agriculture and hydropower generation. Floodplains are intensively managed. Many rivers become fixed in place and locked in time, channelized and canalized to support transport and trade	Cumulative impacts and legacy effects are prominent across much of the planet. Management regimes with a focus on provision of ecosystem services supply and trade-off water to meet agricultural and industrial needs alongside environmental and socio-cultural values. Conservation and restoration projects and designation of environmental flows are localized and expert-driven. Alternatively, a focus on river communities emphasizes living with the river in an ongoing manner, including clean-ups, revegetation and weed maintenance initiatives, and so on.

(continued)

Table 2.1 (continued)

Theme	Hunter-gatherer	Agricultural revolution	Scientific and Industrial revolution	Electronic and Information age
Perceptions of the river and implications for its voice	The lifeblood of the land	A resource to be used	Emphasis on control of floods perceives rivers as drains or sewers that supply water when and where it is needed to support human activities	Approaches to Total Catchment Management apply an ethos of resource provision, including ecosystem services. Alternatively, a living river ethos conceives rivers as part of the web of living things, including humans

supports adaptations to changing land and water conditions, alongside other considerations. Various river gods and guardians express respect and reverence for healthy, living rivers. Human impacts upon rivers are negligible and localized because of low population numbers and densities. Conceived as the lifeblood of the land, rivers are freely able to express their voice, as healthy rivers generate minimal disruptions to societal needs.

Transitions from hunter-gathering activities to herding animals and growing crops changed societal relations to rivers, conceiving water and soil as resources for human use (Table 2.1). In efforts to meet utilitarian functions, sufficient water of appropriate quality was required in the right place at the right time. Significant endeavours made drylands wetter and wetlands drier. Efforts to provide water for others broke many of the relational bonds that characterized day to day relations to rivers. The technical ability of hydraulic civilizations supported extensive water management programmes. Increasingly replicable techniques constructed canals, bridges, water infrastructure facilities and initial variants of sewage treatment plants. Therapeutic, aesthetic and recreational uses of water became more prominent. Agricultural surpluses underpinned the use of rivers to support trade and commerce. The development of boats and bridges, among many factors, transformed the sense of the 'local', as rivers became agents of migration, transfer and trade. Many of the world's largest cities were built around rivers. Many rivers demarcated territorial borders. Healthy rivers served societal interests through provision of resources. Rituals celebrated spiritual connections to rivers. In some ways, however, rivers were increasingly taken for granted; they were only acknowledged when the services upon which society depended were not provided as expected or required. Environmental degradation and climate change triggered the collapse of societies and civilizations, largely in response to limited capacity to proactively address problems.

The scientific and industrial revolutions triggered profound transformations within society and in societal relations to the environment (Table 2.1). Exploitative and materialist relations conceived nature in the service of society, envisaged under religious and/or scientific assertions of human dominance. Engineering prowess provided navigable channels, hydropower and irrigation schemes, flood protection programmes and piped flow networks in urban areas. Direct, purposeful interventions applied command and control practices to tame and train rivers to meet human interests, suppressing inherent variability through the imposition of stable, uniform, hydraulically efficient channels. Despite societal dependence upon rivers, they were treated as sewers and drains, poisoned by chemicals and pollutants, channelized, canalized and dammed, or buried in pipes. Such disdain and the consequences that ensued furthered societal disconnect from watercourses. Rivers no longer presented physical boundaries. Rather, they provided opportunities for economic development in the interests of progress and growth. Healthy rivers delivered water where it was needed, when it was needed, flushing away waste products.

Since the 1960s, a growing environmental movement has sought to improve river health and enhance societal relations to rivers. These initiatives have taken very different forms in different parts of the world (Brierley and Fryirs 2008). As outlined in Table 2.1 and discussed further in Chap. 3, two distinct options have been applied. An economically driven management perspective is framed in terms of provision of ecosystem services, consents processes and environmental trade-offs. This is referred to as a Medean (competitive) approach to river repair. Alternatively, a socio-cultural lens embraces holistic approaches to social and environmental justice, reconceptualizing rivers as the lifeblood of the land, allowing living rivers to express their own voice. This Gaian (cooperative) approach to river repair embraces a more-than-human ethos that seeks to allow each river to express its own voice.

Just as notions of a healthy river have changed markedly over time, they will doubtless change in the future. Stories, relationships and prospects vary markedly from place to place. These are issues of geographic specificity and historical contingency. Much depends upon choices made in shaping approaches to the process of river repair.

References

Ackroyd, P. (2007). *Thames: Sacred River*. New York: Doubleday.
Beattie, J., & Morgan, R. (2017). Engineering Edens on this 'rivered earth'? A review article on water management and hydro-resilience in the British Empire, 1860–1940s. *Environment and History, 23*(1), 39–63.
Best, J. (2019). Anthropogenic stresses on the world's big rivers. *Nature Geoscience, 12,* 7–21.
Bowler, P. J. (1992). *The Fontana History of the Environmental Sciences*. London: Fontana.
Brierley, G. J., & Fryirs, K. A. (Eds.). (2008). *River Futures*. Washington, DC: Island Press.
Budiansky, S. (1995). *Nature's Keepers. The New Science of Nature Management*. London: Weidenfeld & Nicolson.

Capra, F. (1982). *The Turning Point. Science, Society and the Rising Culture*. New York: Simon and Schuster (Bantam Paperback, 1983).

Carson, R. (1962). *Silent Spring*. Boston: Houghton Mifflin Harcourt.

Chakraborty, A., & Chakraborty, S. (2018). Rivers as socioecological landscapes. Chap. 2 in Cooper, M., Chakraborty, A., & Chakraborty, S. (Eds.), *Rivers and Society: Landscapes, Governance and Livelihoods*. London: Routledge.

Chapin, F. S., Carpenter, S. R., Kofinas, G. P., Folke, C., Abel, N., Clark, W. C., … Berkes, F. (2010). Ecosystem stewardship: Sustainability strategies for a rapidly changing planet. *Trends in Ecology & Evolution, 25*(4), 241–249.

Dart, R. A. (1940). Recent discoveries bearing on human history in southern Africa. *The Journal of the Royal Anthropological Institute of Great Britain and Ireland, 70*(1), 13–27.

Dartnell, L. (2018). *Origins. How the Earth Made Us*. London: The Bodley Head.

Diamond, J. M. (1997). *Guns, Germs and Steel: A Short History of Everybody for the Last 13,000 Years*. New York: W.H. Norton.

Diamond, J. (2005). *Collapse: How Societies Choose to Fail or Succeed*. New York: Penguin.

Diamond, J. (2013). *The World Until Yesterday: What Can We Learn from Traditional Societies?* New York: Penguin.

Engels, F. (1910). *Socialism: Utopian and Scientific*. Chicago: Kerr and Co.

Fairbanks, D. J. (2012). *Evolving: The Human Effect and Why It Matters*. Amherst, NY: Prometheus Books.

Flannery, T. F. (2010). *Here on Earth. A Natural History of the Planet*. New York: Atlantic Monthly Press.

Friedman, T. L. (2006). *The World Is Flat: The Globalized World in the Twenty-First Century*. London: Penguin.

Harari, Y. N. (2014). *Sapiens: A Brief History of Humankind*. London: Random House.

Hoefle, S. W. (2016). Além da sociedade-natureza com a mais-que-geografia humana: por uma teoria transdisciplinar de ética ambiental e visão do mundo. In E. S. Sposito, C. A. Silva, J. Sant'anna Neto, & E. S. Melazzo (Eds.), *A diversidade da Geografia brasiliera* (pp. 467–505). Rio de Janeiro: Consequência, UFRJ.

Howitt, R., & Suchet-Pearson, S. (2006). Rethinking the building blocks: Ontological pluralism and the idea of 'management'. *Geografiska Annaler: Series B, Human Geography, 88*(3), 323–335.

Hüttl, R. F., Bens, O., Bismuth, C., Hoechstetter, S., Frede, H.-G., & Kümpel, H.-J. (2016). Introduction: A critical appraisal of major water engineering projects and the need for interdisciplinary approaches. In R. F. Hüttl, O. Bens, C. Bismuth, & S. Hoechstetter (Eds.), *Society-Water-Technology* (pp. 3–9). Cham: Springer.

Jones, J., Börger, L., Tummers, J., Jones, P., Lucas, M., Kerr, J., … Vowles, A. (2019). A comprehensive assessment of stream fragmentation in Great Britain. *Science of the Total Environment, 673*, 756–762.

Macklin, M. G., & Lewin, J. (2019). River stresses in anthropogenic times: Large-scale global patterns and extended environmental timelines. *Progress in Physical Geography: Earth and Environment, 43*(1), 3–23.

Maclean, K., Bark, R., Moggridge, B., Jackson, S., & Pollino, C. (2012). *Ngemba Water Values and Interests: Ngemba Old Mission Billabong and Brewarrina Aboriginal Fish Traps (Baiame's Nguunhu)*. Canberra, ACT: CSIRO.

Marr, A. (2012). *A History of the World*. London: Macmillan.

Marsh, G. P. (1864). *Man and Nature: Or Physical Geography as Modified by Human Action*. New York: Scribner.

McNiven, I., Crouch, J., Richards, T., Sniderman, K., Dolby, N., & Mirring, G. (2015). Phased redevelopment of an ancient Gunditjmara fish trap over the past 800 years: Muldoons Trap Complex, Lake Condah, Southwestern Victoria. *Australian Archaeology, 81*(1), 44–58.

Merchant, C. (1980). *The Death of Nature: Women. Ecology, and the Scientific Revolution*. San Francisco: Harper and Row.

Montgomery, D. R. (2008). Dreams of natural streams. *Science, 319*(5861), 291–292.

Pascoe, B. (2014). *Dark Emu Black Seeds: Agriculture or Accident?* Broome, WA: Magabala Books.

Petts, G. E. (1984). *Impounded Rivers: Perspectives for Ecological Management*. Chichester: Wiley.

Postel, S., & Richter, B. (2003). *Rivers for Life: Managing Water for People and Nature*. Washington, DC: Island Press.

Reisner, M. (1986). *Cadillac Desert: The American West and Its Disappearing Water*. New York: Viking.

Shiva, V. (2002). *Water Wars: Privatization, Pollution, and Profit*. Cambridge, MA: South End Press.

Steffen, W., Grinevald, J., Crutzen, P., & McNeill, J. (2011). The Anthropocene: Conceptual and historical perspectives. *Philosophical Transactions of the Royal Society A: Mathematical, Physical and Engineering Sciences, 369*(1938), 842–867.

White, L. (1967). The historical roots of our ecologic crisis. *Science, 155*(3767), 1203–1207.

Wittfogel, K. A. (1955). Developmental aspects of hydraulic societies. In J. H. Steward (Ed.), *Irrigation Civilizations: A Comparative Study* (pp. 43–52). Washington, DC: Pan American Union.

Wright, R. (2004). *A Short History of Progress*. Toronto, ON: House of Anansi.

CHAPTER 3

Competitive Versus Cooperative Approaches to River Repair

Abstract Contemporary approaches to river repair are economically driven. A competitive (Medean) worldview separates humans from nature, managing rivers as resource and service providers using top-down, command and control practices. In contrast, a cooperative, collaborative, more-than-human (Gaian) approach to living with rivers sees humans as part of nature, conceptualizing the Earth System as a living and emergent superorganism. Bottom-up inclusionary approaches to engagement, participation and governance are framed as holistic and ongoing commitments to place-based, catchment-specific endeavours. Contrasting approaches to conservation planning, restoration activities and co-governance and co-management arrangements are engendered through these alternative framings. Recent river rights legislation offers an intriguing prospect to reframe societal relations to rivers.

Keywords Neoliberalism • Conservation • Restoration • Governance • River rights • River management • Living river

3.1 Scoping River Futures

The future of everything we have accomplished since our intelligence evolved will depend on the wisdom of our actions over the next few years. ... If we fail— if we blow up or degrade the biosphere so it can no longer sustain us—nature

will merely shrug and conclude that letting the apes run the laboratory was fun for a while but in the end a bad idea. Ronald Wright (2004, pp. 3, 31)

Although no-one would deny the importance of healthy environments, approaches to achieve and maintain this are highly contested socio-political issues. Healthy rivers may well be products of healthy societies, but what kind of society does that reflect?

Notable improvements to river health have been achieved across many parts of the world in recent decades. Growing environmental consciousness and alarm for degraded environmental conditions have triggered an era of river repair (Brierley and Fryirs 2008). The timing of this movement has coincided with political, socio-economic and cultural drivers of a neoliberal era of hands-off government in the western world. In this period, sources of funding in the environmental sector have shifted from government-sponsored programmes to the world of private capital, often orchestrated through market-driven mechanisms based on compensation or mitigation (Lave et al. 2008; Robertson 2006). Following Flannery (2010), such framings are referred to here as a Medean lens—the user-pays workings of a competitive, dog-eat-dog, consents-based, limit-setting world.

This chapter contends that moves towards an environmentally framed agenda have failed to reach their potential because the influence of an alternative green movement has been systematically undermined by neoliberal imperatives, inhibiting prospective uptake of a cooperative, Gaian approach to river repair that emphasizes closely woven linkages between environmental and human health (after Lovelock 2000; see Flannery 2010). Gaian framings focus upon societal relations and interactions that seek to improve the health of ALL rivers, living with living rivers rather than seeking to manage them to a particular image. This is NOT a fix-and-forget activity or a project. Rather, an ongoing commitment focusses upon how we live with the world and each other (see Box 3.1).

Box 3.1 Contemplating Prospects for River Repair in India
Sharing river stories is fun. A few years ago, I met an engineering student from India at Vancouver airport. A recent trip to the Kosi River in Bihar had left a profound impression of societal disconnect from a sacred river system, so I said to this fellow, "Just what are you going to do to look after the Ganga, your Mother River?" His response was instantaneous: "Why do you think we invest so much in research on Mars?"

Although the Anthropocene originated as a geological description of a new period in Earth history, it is by necessity an ethical concept that embeds various interconnected social, economic and ecological contexts (Kersten 2013). In this 'Age of Humans', geological strata include industrial emissions, nuclear signatures and pollen from industrial agriculture, but exclude fossils of species driven to extinction by human activities. While the underlying facts of the Anthropocene are scientific, the term conveys a message of almost unparalleled moral-political urgency under the guise of scientific neutrality (Sloterdijk, quoted in Davison 2019). Implicitly, assertions of the Anthropocene draw together multiple crises of our times as a platform to put a label on the issues that must be addressed, thereby presenting a starting point to develop and apply measures to solve them (Purdy 2015).

Table 3.1 presents an overview of the mindsets and frames of reference that underpin a competitive (Medean) and a cooperative (Gaian) approach to river repair. Inevitably, this binary juxtaposition of contested worldviews is a gross simplification of reality, as conflicting elements and drivers sit alongside each other in differing combinations in differing situations in space and time. However, the distinctions outlined in Table 3.1 provide a basis to differentiate among prevailing premises, approaches and practices in framing endeavours in the era of river repair, and the outcomes and trajectories that ensue.

According to Flannery (2010), the Medea hypothesis represents a genetic predisposition for self-destruction, while the Gaia hypothesis perceives the world as an interconnected superorganism in which living with each other presents a preferable outcome relative to killing each other off. Medea is a sorceress from Greek mythology, descended from the sun god. A Medean world is exploitative and elitist/exclusionary. A competitive ethos separates humans from nature, perceiving nature as something that has been constructed to serve human interests. The environment is considered to be controllable when subjected to human will. A functional, utilitarian perspective views nature as a provider of services and resources to be exploited and managed. An anthropocentric and materialist worldview envisages people in charge of rivers, imposing measures to conquer, tame and train them. Command and control measures envisage rivers as machines. Engineering practices make rivers the same, applying fix and forget practices that view nature as 'out there'—something that can be put aside. Sacrificing river health to achieve social and economic goals is a

Table 3.1 Approaches to river repair through a Medean and a Gaian lens

Principle	Realities of the contemporary Medean world—the world as it is	The world as it could be: A Gaian approach to river repair
Mindset and mentality that frames societal relations to rivers		
Value set	Anthropocentric, utilitarian and materialistic approaches to river management focus upon economic principles, market forces, ecosystem services and consents-based approaches to mitigation, limit-setting and trade-offs	River- and eco-centric, harmonious, more-than-human—a holistic approach to living with living rivers
Perspective upon the river	An entity that is 'out there', to be managed as 'norms' or 'types'. Perceives rivers as predictable entities. Endeavours to quash uncertainty. Command and control practices fix channels in place and lock them in time, creating uniform and tidy rivers. Form-based measures emphasize aesthetics over functionality	Embraces uncertainty as an integral component of a living river ethos, emphasizing the coevolution and mutual interdependence of biotic, abiotic and socio-cultural attributes of each and every river. Lives with the diversity, variability and uncertainty of dynamically adjusting and evolving systems. Catchment-specific applications focus attention upon things that matter, applying holistic principles in a whole of system approach. Rivers are complex, often quite messy entities
Knowledge base and approach to learning and communication		
Forms of knowledge	Theoretically framed, scientific and technical knowledge emphasize concerns for stability and certainty. Conceives rivers as bits or components, in spatial and disciplinary terms	Empirical, place-based understandings apply a multiple knowledges ethos, striving to tell each river's story. Merges scientific and local knowledges, including stories, narratives, and so on.
Application of knowledge	Expert-driven, deterministic applications impose external understandings, typically framed within a techno-fix mentality	Collective approach to learning through shared commitment to experimentation

(*continued*)

Table 3.1 (continued)

Principle	Realities of the contemporary Medean world—the world as it is	The world as it could be: A Gaian approach to river repair
Monitoring	Externally framed and applied	Combines external applications with local initiatives, such as citizen science and community monitoring programmes

Approach to managing the process of river repair

Power relations, decision-making processes and prioritization	Economy first and foremost: A world of ecosystem services, limit-setting, trade-offs, consents processes and negotiated (compromise) solutions. Issues-based practices manage parts of the system, one problem at a time. Generic top-down practices are imposed upon local situations	River first: The economy is a wholly owned subsidiary of the environment. Endeavours to 'Find the Voice of the River' allowing the river to speak for itself. Bottom-up and top-down perspectives are merged through inclusive, locally owned practices that scale-up effectively within coherent visions and policy framings
Approach to environmental protection	Protection zones and reserves are separated from human activities in a zoo-like mentality that views nature as 'out-there' (behind fences) while humans are 'over-here'	An ecosystem approach to living with all rivers applies a conservation-first ethos that looks after good bits and things that matter, before they become a problem
Approaches to restoration practice	Emphasis is given to 'doing things', as that's where the money lies. Reach-scale applications apply consistent and standardized measures that manage rivers as particular 'types', making rivers the same	Place-based (catchment-specific) measures leave the river alone as much as practicable, allowing it to look after itself. Passive action, the 'do-nothing' option, is a strategic decision in its own right. Tailored practices fit local circumstances, giving due regard for off-site impacts, proactively addressing threatening processes
Approach to implementation	Externally applied practices are designed and implemented by consultants and contractors (expert-based, employing professional practitioners). The restoration 'industry' emphasizes the quest for profit	Locally implemented community-led practices build upon societal relations to the river (living with the river), engaging collectively in the process of river repair through hands-on, grassroots endeavours

bottom line reality of a limit-setting, consents-based world. Only the fittest survive in this world of ruthless selfishness, while the weak and poorly adapted simply die out as evolution progresses (Flannery 2010). There is every prospect that the final step in conquering nature will be a self-defeating enterprise for humanity too. Economic growth cannot occur at the expense of the environment indefinitely. However, as noted by Dawkins (2017, p. 58): "In an evolutionary world of fundamentally selfish entities, those individuals that cooperate turn out to be surprisingly likely to prosper."

Gaia is the Greek Earth Goddess. A Gaian world is collaborative, cooperative and inclusive, viewing nature as a living and evolving organism (Lovelock 2000). A complex web of interactions links all aspects of the earth's physical structure and the living things that inhabit it. Recognizing the fundamental importance of principles of mutual interdependence, coevolution and reciprocity, efforts are made to look after living and emergent rivers as they look after human society. This holistic and ecological conception of the world sees the universe as a living system comprised of interrelated and interdependent biological, psychological, social and environmental phenomena. A Gaian ethos entails concerns for how we live with all rivers. Differentiation of the economy from the environment is viewed as a false dichotomy, perceiving the economy as a wholly owned subsidiary of the environment. Habits of mind consciously apply a duty of care that is inherently place-based on the one hand, yet innately global on the other.

Transitioning from a Medean to a Gaian world is exceptionally challenging. Inertial forces, vested interests and change resistance support the maintenance of the status quo. However, contemporary practices are not working. Environmental crises imperil our future. Existing governance arrangements are simply unable to cope with the demands of our time. Pushing aside Medean assertions of challenge and drama, Gaian attributes of harmony, warmth and affection offer a more wholesome prospect for societal and environmental wellbeing.

Approaches to managing or living with rivers have profound implications for conservation, restoration, governance arrangements and assertions of river rights (Table 3.1). Prior to addressing these themes, conceptual underpinnings of neoliberalism that underpin machinations of the contemporary Medean world are briefly outlined.

3.2   NEOLIBERAL APPROACHES TO ENVIRONMENTAL MANAGEMENT

[O]f all the forms of pollution, that of the *mind* was most serious. Roderick Frazier Nash (2001, pp. 389–390)

Profound societal and political adjustments took place following the ravages of the Second World War and the preceding Great Depression. State-sponsored development programmes included various social welfare, health and infrastructure initiatives, as a blend of state, market and democratic institutions sought to create full employment and economic growth while addressing concerns for the welfare of citizenry (Harvey 2007). However, since the 1970s, neoliberal policies have seen the deregulation, privatization and withdrawal of the state from many areas of social provision. Private economic agendas driven by market forces have become *de rigeur*. Globalizing influences have channelled increased corporate power and wealth from subordinate classes to dominant ones and from poorer to richer countries, systematically dismantling institutions that promoted egalitarian distributive measures in the preceding era (Harvey 2007).

Neoliberalism is a theory of political economic practices that proposes that human wellbeing can best be advanced by liberating individual entrepreneurial freedoms and skills within an institutional framework characterized by strong private property rights, free markets and free trade (Harvey 2007). The state creates and helps to maintain an institutional framework that supports such practices, securing private property rights and guaranteeing, by force if need be, the proper functioning of markets, helping to establish new market mechanisms if required. The market-driven, user-pays world of limit-setting, consents processes, reduced taxes and self-regulation has transformed the delivery of services in healthcare, education, social security and environmental management. The hollowing out of the role of the state has created a niche market for private consultants and operators. At an individual level, there is money to be made in applying economic measures to mitigate and remedy damage, rather than avoiding it in the first place. In the long term, society pays.

Although it does not make sense to sell nature to save it (Dempsey 2016), monetarist perspectives have become *de rigeur* in neoliberal approaches to environmental management (Table 3.1). Indeed, valuing nature has become a business in itself, as environmental assets, resources and reserves are bought and traded as commodities (Lave 2018). Risk

assessment procedures and cost-effective practices incorporate environmental values into operational models, protecting, managing and marketing particular assets and ecosystem services (Jackson and Palmer 2015). In
the corporate world of big business, concerns for environmental justice, if
they exist at all, are strictly subservient to concerns for economic efficiency, continuous growth and capital accumulation (Harvey 1996,
p. 375). Strong policies and monitoring programmes for Clean Air or
Clean Water do not serve the interests of powerful companies in the
resource and agricultural sector. Hence, formal commitments to monitoring and maintenance have been pushed aside in a world of short-term
project-based contracts, as fix and forget measures seemingly assume that
problems disappear if they are not measured, reported upon and publicized.

Across much of the world, the idea that fresh water is part of the commons is under siege. While state-based mechanisms had created infrastructure to supply water and build sewage treatment facilities, deregulation
saw the selling-off of assets and moves towards privately run demand management framings. More money was to be made through greater efficiency of water use and associated market-driven pricing and trading
mechanisms. Commodification of water, treating it as an object or a product in the service of society, conveys an anthropocentric, utilitarian vision
that fosters the long-standing illusion that people are in charge of the
planet (Salmond 2017).

Concerns for water security and the global water crisis are not issues of
scarcity; rather, they reflect a failure of governance (Cook and Bakker
2012). Increasingly, concerns for water management as a basis of life for
all living things are set aside through programmes that serve the benefits
of a few. All too often, decisions over water allocations reflect the politics
of power and influence rather than the public good. Concerns for property, ownership and land tenure arrangements impede prospects for collective actions at broader scales. In the collaborative world of a Gaian lens,
the Earth belongs to no one. Shared commitment to cooperative management of the commons applies a sustainability lens to maintain living and
adapting systems, taking care to negate exploitative practices (cf., Hardin
1968; Ostrom 2009).

In the Medean world of neoliberal practices, mitigating mechanisms
include trade-offs for pollution, water, carbon and ecosystem services.
Contestations around values abound in the trade-off between pricing
mechanisms and economic incentives, development pathways for industries
and agriculture, ecological conservation/rehabilitation, recreationalists,

spiritual-cultural connections and indigenous rights to water. Offset and mitigation programmes transfer environmental benefits by recreating habitat in other places to compensate for damage in situ (Lave 2018). The market-driven world of offsets, trade-offs and stream mitigation banking has created countless initiatives to reforest landscapes, regenerate grasslands and reconstruct wetlands, typically in areas where past land management practices cleared forests, over-grazed grasslands and drained swamps. A pioneering spirit of re-making rivers has become big business, especially in the United States (Lave 2018). Indeed, wetland (re)construction has become a significant industry in some areas, making wetlands available to developers as credits to meet planning requirements (Eden 2017). Such transfers of equivalent value (monetary and ecological) create and trade habitat in the same way that commodities such as food or cars are considered. Parcels of land, presence of species or ecological functions are commodified and used to justify development in the interests of capital. Drawing on political economy, such practices create and substitute forms of natural capital, perceiving nature as something that can be moved and replaced elsewhere (Eden 2017).

Commodification and commercialization of a river, turning it into a collection of products (Strang 2004), separates a river from its broader context and meaning, creating instruments to measure and price it, while unpicking its essential metaphysical and spiritual connections (Muru-Lanning 2016). What price for social rejuvenation and/or psychological wellbeing and cultural values? Can you get a discount? Limit-setting framings and consents processes that treat land and water as property regulate how much pollution or degradation of nature can occur within the law, giving no standing to nature itself and associated non-human elements. Such binary oppositions split mind and matter, the environment from people, the natural from the social sciences, and people from the communities and rivers that sustain them, engendering habits of mind that suggest choices must be made between economic prosperity and the environment (Salmond 2014). While a negotiated deal among stakeholders may make sense in monetary terms, compromise solutions may be nonsense in environmental terms. After all, what is half a habitat? Life cycles are sustained and complete or they fail. Ecosystems are viable or they are not; dead or alive.

Negotiated trade-offs and compromise solutions result in winners and losers, improving the health of some rivers at the expense of others. Environmental interests are the big losers in the compromise world of lowest common denominator practices. The use of economic trade-offs to optimize values to differentiate among allocation scenarios reduces the plurality of human experience to a small number of categories to be optimized (Tadaki and Sinner 2014). Such practices discount the future, reckoning that if you are unlikely to see tomorrow, you may as well take what you can get today, even if it means foregoing a far greater reward in the future (Flannery 2010). The rate of environmental degradation may be locally reduced, but it will not be reversed, as things get worse more slowly, rather than improving.

Concerns for human welfare and the wellbeing of society extend well beyond the marketplace. There is an inherent contradiction in placing a finite value on an irreplaceable life-support system, whereby the more scarce the ecosystem services, the greater their 'value' (Postel and Richter 2003). Such framings dilute prospects for genuine and authentic engagement in the process of environmental repair.

3.3   KNOWLEDGE-BASES AND PRACTICES THAT UNDERPIN APPROACHES TO RIVER REPAIR IN A MEDEAN AND A GAIAN WORLD

You can't part a great river, because it is both greater and other than its parts: its constancy and immensity of flow are a *union*, the antithesis of parts. David James Duncan (2001, p. 6)

In a Medean world, anthropocentric realities are imposed upon a river through technically focussed applications implemented through top-down practices (Table 3.1). Historically, such interventions have been imposed by governments, but increasingly they reflect private interests. All too often, expert knowledges of theoretical and modelled realities emphasize concerns for stability and certainty. Prescriptive applications of reductionist, discipline-bound knowledges assume researchers know and can provide all the answers. Rivers are conceptualized, analysed and managed as collections of components, typically with little concern for consultation let alone involvement with wider society. Monitoring is externalized, typically minimalized to the stipulated requirements of statutory rules in a self-regulating context.

A Medean approach to river restoration entails negotiations among divergent values and goals, appraising what a river should look like and how it should work (Emery et al. 2013). Restoration programmes typically strive to attain the best achievable state and functionality under prevailing and likely future conditions (e.g. Harris and Heathwaite 2012). Unfortunately, many restoration projects fail to stipulate specific goals, making it exceedingly difficult to assess the effectiveness of practices in economic or environmental terms (Jähnig et al. 2011; Palmer et al. 2005).

A Gaian world lives with the diversity, variability and uncertainty of living, dynamically adjusting rivers, perpetuating the coevolution and interdependence of biotic, abiotic and social attributes. Knowledge generation and use emphasize innately holistic framings, focussing upon the interconnectedness of all working parts, linked through symbiotic and evolving interactions (Table 3.1). A multiple methods, multiple lines of evidence approach derives conceptual models that assess the operation of the system as a whole. Place-based applications incorporate local values in assessing 'things that matter', seeking to protect and/or enhance such attributes and relationships. Constructive and reflective approaches to ongoing learning apply open-ended, non-prescriptive thinking, embracing uncertainty through a commitment to experimental practices, creativity and innovation. Community-based monitoring programmes and citizen-science initiatives adapt practices as required. Concerns for collective understandings and actions incorporate emotional and psychological considerations, framing informal knowledges alongside scientific and technological prowess. An ethical commitment to social and environmental justice builds upon habits of mind—lifestyle choices that shape societal relations to rivers.

In his book entitled *The Turning Point*, Capra (1982) envisaged the inevitable emergence and adoption of a holistic ecological worldview. While the prospect and potential are good, such framings are yet to become a day to day reality. As powerful vested interests are well served by existing political, institutional and governance arrangements, it is in their interests to manipulate and accentuate change resistance to maintain the status quo, rather than engaging more generatively in efforts that serve the collective interests of public good.

Discipline-bound science and technical know-how played critical roles in the exploitative ventures that asserted human authority over Planet Earth. Since the Scientific Revolution, professional fragmentation of science in an age of specialization, reductionism and discipline-bound enquiry came to symbolize the material trend to divide Nature into separate units, "each of which can be studied in isolation and exploited for short-term profit" (Bowler 1992, p. xiv). However, such siloed thinking and application does not provide an appropriate platform for the era of river repair. All too often, fractured framings, knowledge structures and contested relations between disciplinary experts engender fragmented management practices and outcomes. Recurrent quests for synthesis in earth and environmental sciences have been pushed aside as an unworkable idealism. Building upon foundation work in natural history conducted by Alexander von Humboldt (1769–1859), the emergence of ecology provided a stepping stone for developments in environmental science. In his original use of the term Oecologie, Haeckel (1866) denoted the study of the interactions between organisms and the external world (Bowler 1992). Such assertions related a sense of unity of Nature to a sense of home, a form of global organic economy. An ecosystem approach came to emphasize the study of the interactions upon which the whole network depended (i.e. its functionality). In a related manner, Penck (1897) made a plea for the conduct of integrative river science under the label potamology, combining principles from hydrology, geomorphology, ecology, chemistry, engineering and related fields.

Efforts to work within disciplinary silos then stitch things together as an integrative whole do not work. Holistic ecological knowledge seeks to reintegrate the biophysical sciences into the human sciences to develop understandings of the inherent complexity of socio-environmental issues, re-conceptualizing humans as part of a much larger organismic system (Capra 1982). Rather than being exploitative and dominating, holistic framings provide a critique of capitalist developmental models. Inclusive approaches to knowledge generation unite local and scientific knowledges to support collective and participatory action. Broader conceptualizations of 'culture' and 'environment' embrace and accept the possibility of a spiritual dimension. Coordinated and visionary practices interweave socioeconomic and cultural values as integral components of environmental programmes. Local and traditional knowledges are framed alongside scientific insights in applications of a holistic, more-than-human lens that generates place-based understandings through a multiple knowledges ethos. An emphasis upon things that matter engages local citizenry in generative steps to live with their river in an ongoing commitment to healthy, mutually interdependent relationships—with the environment and with each other. Rather than focussing upon 'what can we afford to do?', emphasis is placed on 'what can we not afford not to do'?

Ways of knowing and approaches to enquiry are socially situated and structurally constrained. Contestations abound in negotiating the gulf between informal understandings, typically applied within a trial and error approach, and the formal rigours of rational scientific discourse, prescribed through experimental procedures that apply a particular teleological (purposeful) worldview. All too often, a posture of moral neutrality reinforces and protects the value parameters of the status quo, as rules of engagement inhibit discussion let alone uptake of alternative strategies (Meppem and Bourke 1999). In these highly politicized activities, power relations and gatekeepers of knowledge protect vested interests, privileging and embedding particular ways of knowing. This shapes the work that is done to the exclusion of alternative ideas and approaches, delegitimizing other ways of knowing and their associated insights. Adopted procedures and the results they generate reflect, and are used to endorse, the values of those who created them, in what Haraway (1991) refers to as the God-trick.

Scientists like to think of themselves as the exponents of an objective method designed to generate factual knowledge about the world. However, notions of rationality and truth based on purely objective standards convey a mischievous falsehood, failing to recognize external influences. Truth is to science what faith is to religion. It is not, and cannot be, absolute, yet believers sometimes contend that it may be. True science is never philosophically partisan—a disembodied view-from-nowhere (Shapin 1998). Rather, scientific knowledges are shaped by the social and material environments within which they are produced. As postulated by Worster (1993), science can be conceived as 'culture and politics by another name' while Budiansky (1995) asserts that science is a tool, not a moral imperative. Particular models of nature are designed to legitimize particular sets of social values.

It is only in recent decades that science has been conducted in the interest of social and environmental justice. Perhaps inevitably, the environmental movement frames scientific endeavours in quite different ways to those projected by the military-industrial establishment. In such deeply contested spaces, technocentric framings are imposed upon society by governments (notionally in the interests of society) and corporations (in the interests of profit), asserting various forms of authority through the appointment of particular gatekeepers of knowledge (Box 3.2).

Box 3.2 Contestations in the Generation and Use of Knowledge
Appointed gatekeepers constrain the production and dissemination of knowledge. For example, publishers determine which book contracts are awarded, while journal editors guide the peer review process through selection of referees, asserting control over what gets published. In a truly bizarre suite of arrangements, governments subsidize a significant part of research production, yet owners of the journals who publish findings engage reviewers at no (or negligible) cost, hold the copyright for publications, impose fees to produce journal articles in an open-source world, take limited risks and walk away with the profits. Graduate committees appoint PhD examiners to guide rights of passage, facetiously referred to as a muppet and a mate. Particular processes appoint leaders or ascribe membership to professional bodies and networks, shaping accreditation programmes and opportunities for professional training. Such practices frame the codes of conduct that influence the research that is funded and the questions that are asked (by whom), approaches to information/ data gathering and analysis, and approaches to monitoring and communication. Commercialization and corporatization of knowledge have markedly blurred the boundaries between public-good research and consulting services conducted through the private sector. Increasingly, the role of universities as the critic and conscience of society is open to question. Marginalization and alienation reflect the everyday injustices of neglect.

The ever-expanding range of datasets, modelling applications and toolkits in the armoury present an unprecedented capacity to develop and apply catchment-specific understandings (Fryirs et al. 2019). Toolkits and modelling frameworks not only represent the world, they transform it, embedding particular sets of values to the exclusion of others, whether consciously or otherwise (Lane 2014; Lave 2016; Tadaki et al. 2014). Many tools and models are not necessarily designed to be used collectively to serve the greater good. Rather, commercial interests come to the fore, masking inbuilt assumptions and limitations, limiting prospects for verification unless a fee is paid. Indeed, the language used in such deliberations, the tools, tool-belts and tool-kits in the armoury, smack of materialism and functionalism of a command and control ethos applied under the auspices of the military-industrial complex! How ready are we to replace the perceived certainty of engineering-based approaches to uniformity, stability and predictability with the inherent uncertainty that accompanies place-based analyses of diversity, variability and evolutionary trajectory (Hillman and Brierley 2008)? Such determinations have major implications for approaches to environmental conservation, restoration and governance arrangements.

3.4   MEDEAN AND GAIAN APPROACHES TO ENVIRONMENTAL CONSERVATION

Conservation is getting nowhere because it is incompatible with our Abrahamic concept of land. We abuse land because we regard it as a commodity belonging to us. When we see land as a community to which we belong, we may begin to use it with love and respect. … That land is a community is the basic concept of ecology, but that land is to be loved and respected is an extension of ethics. Aldo Leopold (1949, reprinted 1968), Foreword, p. viii

What options are available to conserve landscapes and ecosystems? The Medean option envisages management of wilderness values in Environmental Protection Zones, externalizing people from nature by fencing off reserves, wilderness zones or national parks. In a sense, such areas are managed as large scale zoos with restricted and exclusive access to practitioners with fishing, hunting and gaming licences or wealthy eco-tourists. Why put yourself through the angst and turmoil of stakeholder negotiations as part of participatory practices in the quest for environmental solutions when you can simply purchase an environmental asset such as an area of land to create or add to a reserve or buy/trade a carbon credit, through procedures that are efficiently orchestrated by an intermediary? Such is the dilemma of contemporary environmental practices and obligations. Acts of environmental vandalism in other places can be traded-off and negotiated out of existence by a sponsorship logo on an environmental reserve or a fenced-off protection area with restricted access to most of society. Medean approaches to environmental protection are riddled with inherent hypocrisies and contradictions (Box 3.3). In some ways, such conservation programmes are more realistically conceived as the creation of a museum exhibition or the machinations of a virtual world. In contrast, the Gaian option incorporates conservation principles and practices into day to day activities, living with adjusting and emergent riverscapes and ecosystems.

Box 3.3 Some Perversities of Conservation Management
Numerous environmental visionaries set out to protect wetlands, old growth forests and other habitat refuges. Unfortunately, many hard won conservation gains are easily lost as legislative changes support the interests of development. Several contradictions and perversities of the conservation movement are outlined below:

- Early conservation efforts primarily served the hunting interests of the wealthy, in a similar manner to trophy hunting game reserves in Africa. Some quite perverse outcomes have been created. For example, the first national parks in New South Wales, Australia were established to protect deer, an introduced species. In New Zealand, fish species introduced by acclimatization societies as a food supply and to support recreational activities are the only named aquatic species protected in legislation (Blue 2018).
- Landscapes in various parts of the United Kingdom have been devoid of trees for so many generations that re-establishing forest cover is popularly viewed as destroying the landscape patrimony (Monbiot 2013). Similarly, following the death of King Valdemar the Young in a hunting accident in 1231, trees are no longer allowed to grow in an area set aside as a reserve on the Roesnaes Peninsula near Kalundborg in Denmark.
- Some of the most precious remnant rivers in coastal catchments of New South Wales, Australia, are in areas where landowners discouraged government officials from working on their land back in the 1970s. The prevailing management practices of the time imposed a controlling and stabilizing ethos using rip rap, gabions and exotic vegetation. As these landowners were 'difficult', they were left alone. So were their rivers. Although these landowners were considered to be rebels at the time, they are now considered to be visionaries.
- Many assertions of wilderness are undeniably exclusionary. For example, William Wordsworth demonstrated an elitist streak in a letter penned in 1844, bemoaning prospects that a proposed railway would bring trainloads of untutored working class sightseers from the industrial northwest to his beloved Lake District in England (quoted in Budiansky 1995, p. 29).

- Removal of traditional people from their lands as ecological migrants is especially ironic in the environmental protection programme that underpins the Sanjiangyuan National Nature Reserve in western China, as the grazing-adapted ecosystems to be protected in this area are products of traditional land use practices.
- Many of the areas for which Wild and Scenic River status has been sought are populated by indigenous peoples remote from major population centres. Under rights of self-autonomy, such peoples do not want their lifestyle options to be prescribed by others (e.g. Howitt 2018). For example, when the Wild Rivers Act was passed in Australia in 2005 to preserve natural heritage, Aboriginal leaders condemned the bill as it restricts their rights and overrides their ability to use the land around the river for commercial purposes (Slater 2013).

Few would deny the importance of biodiversity management and conservation initiatives as part of environmental protection programmes. Preservation of heritage and biodiversity values is often restricted to a world of remnants, many of which are remote, isolated and inaccessible, located in areas that offered limited prospect for exploitative human endeavours. All too often, conservation practices and outcomes reflect poorly controlled experiments conducted in the face of bad policy and bad or inadequate science (Budiansky 1995). Most conservation areas are too small to cover the range of keystone species that they seek to protect. Corridors are required to link them, supporting migration pathways. Fragmentation problems are compounded by the isolation of many reserves. If maintenance of viable populations is the aim, a diverse gene pool is required to avoid inbreeding. In some instances, conservation programmes require culling or control of predators or captive breeding and reintroduction. Recolonization and routine interchange between isolated populations may be vital to maintain genetic diversity. How can particular habitats and species be prioritized in efforts to save and maintain conservation hotspots?

By the late nineteenth century, a growing tension had emerged between those who wanted unrestrained exploitation of nature, those who wanted controlled (sustainable) exploitation to protect future resources and a vocal minority who wanted to protect natural values of wilderness areas (Bowler 1992, p. 505). Various national parks created in the western United States by the end of the century prompted world-wide moves towards protection of the conservation estate. Early proponents of the conservation movement sought to preserve refuges of aesthetic beauty in areas that couldn't be used for other things. However unrepresentative and unintentional this may be, these areas came to instil nostalgic sentiments that preserved particular myths about the land and human relations to it.

Wilderness is a social construct, a culturally and historically contingent expression of anthropocentric, utilitarian and elitist ways of seeing nature (Cronon 1996). A wilderness lens engenders a sense of landscapes that is remote, exotic and out there—unpeopled and separate from human endeavour. This particular ordering of nature views the subject as a disembodied and distanced observer. For much of human history, the wild and untamed world was an unrecognized and un-named norm. Once wilderness was created as a concept by civilization, it was widely hated and feared (Budiansky 1995). In recent centuries, however, wilderness became appreciated and even venerated, as romantic assertions of pristine landscapes and the aesthetics of the sublime came to the fore. Over time, wild places have come to be associated with exclusivity, supporting lifestyle values of a privileged few (Nash 2001; Box 3.3).

In recent decades, large scale conservation programmes have been supported by NGOs and private concerns, such as The Nature Conservancy, Ducks Unlimited, Trout Unlimited and many other corporate and philanthropic ventures (Collard et al. 2015). Such groups buy up large areas of land to protect fishing and hunting rights, and associated river and wetland values. As these large estates have a single owner, they do not encounter problems associated with the Tragedy of the Commons (Hardin 1968), wherein no-one takes responsibility for collective interests. However, there is a certain perversity in perceiving nature as something that is 'out there', applying a zoo-like mentality of fenced reserves that keeps people out and species in. Increasingly, efforts to manage ecosystems can be considered analogous to 'Playing God', as management programmes determine an appropriate mix of species and processes for a given setting.

An ecosystem approach to environmental management is applied at the landscape scale, recognizing that ecosystem processes are emergent, new species mixes and evolutionary traits are inevitable and there are too many species to save them one at a time. Trying to conserve and protect a non-stationary entity is an impossible task. Notions of rewilding convey an intriguing alternative to conventional approaches to environmental conservation (Monbiot 2013). Rewilding is about humility, about step-ping back. The idea is not to control, protect or restore ecosystems so much as to allow ecological processes to take their own shape. Restoring trophic diversity and the role of keystone species enhances opportunities for animals, plants and other creatures to feed on each other, rebuilding the broken strands of the web of life. Unlike conservation, rewilding has no fixed objective. It is driven not by human management but by ecologi-cal processes themselves. There is no point at which it can be said to have arrived—the process itself is the outcome. Monbiot (2013) outlines how the reintroduction of wolves to Yellowstone Park in the United States resulted in the rapid reforestation of the riverbanks and marked increases in beavers, songbirds, otters, muskrats, fish, frogs, reptiles and even bears. Such is the inherent diversity and dynamism of a self-healing and emer-gent ecosystem conceptualized through a Gaian lens.

3.5   Medean and Gaian Approaches to River Repair

There is something remarkable about the way people connect with place and each other through ecological restoration ... When ecosystems come together, so do we ... Restoration satisfies emotional and moral hungers, too. We experience a lift from knowing that our depredations are at least partially absolved through restoration. We are brought to the brink of humility when we realize how easy it is to act carelessly and how hard it is to rebuild and restore. ... restored ecosystems teach us a great deal about ourselves. Eric Higgs (2011, pp. xvii–xix)

Restoration entails key decisions regarding the forms, processes, ecol-ogy and aesthetics of a river. Socio-economic, cultural and political consid-erations determine what is desirable and/or acceptable. In these highly politicized activities, particular perspectives and practices are selectively applied over others.

In a Medean world, attempts to restore nature reflect human arrogance in assuming the natural world can somehow be 'fixed' (Table 3.1). This conceitedness is even more pointed as damage was caused by human activ-ities in the first place (Eden 2017). Much of the work conducted under the auspices of environmental restoration is inauthentic and fraudulent, malicious even (e.g. Katz 1996; Light 2000), 'Faking Nature' through a replacement thesis (Elliot 1997) that trades off concerns for one place with another. Such framings give precedence to the interests of development over environmental protection in a relabelled or rebranded expression of human authority over the river.

In some ways, terminology and definitions are symptomatic of the issues to be addressed. The term 'restoration' has become accepted as a focal point for the process of river repair. Explicitly, this term focuses on the past. As a 're-' word, it not only uses past conditions as a basis to guide contemporary aspirations, it also (re)applies existing framings and practices within a different frame of (re)ference, (re)asserting anthropocentric interests upon the river. Such a Medean mentality continues to break the world up into manageable bits. It is expert-driven, giving primacy to technical knowledge in discipline-bound applications that now incorporate concerns for environmental values within 'soft' or 'sensitive' engineering practices.

In environmental terms, restoration does not make sense. Landscapes and ecosystems evolve. They are emergent, never going back in precisely the same way. In this sense, restoration is an entirely valid concept in returning an art work or a heritage building or a cultural icon to its former glory, but its application to rivers, landscapes and ecosystems is questionable. As it is not possible to restore a living entity, is the wrong message being conveyed? Emphasis should be placed on the world as it could be—as it is becoming—rather than the world as it was. Recognizing that ignorance of the past dooms us to repeat it, lessons from the past should guide our efforts, not least in our endeavours to stop doing stupid things. As noted by William Shakespeare in The Tempest, "What's past is prologue".

A step by step approach to restoration practice, in which each step entails expert application of discipline-bound knowledge, does not work. Yet this is exactly what has happened. Competitions between disciplines create fractured relationships and outcomes. Initial moves to improve river health addressed concerns for water quality. Subsequently, the river health debate focussed attention on concerns for biodiversity values, focussing upon the integrity, vitality and functionality of aquatic ecosystems. Concerted efforts to improve flow conditions along regulated rivers brought about the allocation of environmental flows (see Box 3.4). It was only towards the end of the twentieth century that emphasis was given to concerns for the dynamic physical habitat mosaic of river systems. This brought about a transition beyond a singular concern for channel stability, hydraulic efficiency and hazard mitigation towards a more holistic focus on concerns for river health, biodiversity and sustainable management (Downs and Gregory 2004; Hillman and Brierley 2005). Initially such restoration activities were conducted under the aspirational auspices of the field of dreams hypothesis—create the habitat and the species will come. Unfortunately, this principle did not work, as limiting factors came into effect, wherein the least effectively functioning attribute of the system determined the performance and vitality of the system as a whole (Hilderbrand et al. 2005). As a result, most small-scale restoration practices failed to achieve notable and sustainable improvements to river health (Bernhardt et al. 2005; Collier 2017; Palmer et al. 2014).

Box 3.4 Use of Environmental Flows to Help Silenced Rivers Find Their Voice

Managing the quantity, timing and quality of flow releases below a dam seeks to address concerns for continuity of flow along fragmented (or silenced) rivers. Environmental flows also enhance water quality, regenerate physical habitat, support the life-cycle needs of fish and wildlife and improve the livelihoods of river-based communities. Alongside ecological and environmental benefits, flows increasingly support recreational and socio-cultural purposes, such as boating, aesthetic and cultural flows.

There are two main types of environmental flow (Arthington 2012). First, available flow is manipulated to replicate the natural flow regime in efforts to meet objectives relating to biodiversity and ecological integrity (Poff et al. 1997). Increasingly, such applications incorporate understandings of non-stationarity in climate, environmental conditions (temperature, sediment, nutrients) and ecological features (non-native species spread), moving beyond static, regime-based flow metrics based solely on analysis of past reference conditions to incorporate dynamic, time-varying flow characterizations (Poff 2018). Second, designer flows endeavour to achieve specific ecological and ecosystem service outcomes while meeting human needs for hydropower generation, irrigated water supply and transportation requirements (Acreman 2016). For example, high flow periods can support channel maintenance and floodplain connections while low flows of sufficient duration and correct timing can permit fish migration and spawning (e.g. Jowett and Biggs 2009; Olden et al. 2014). Determination of the 'desired ecosystem' is key to such deliberations (Acreman 2016).

Designation of environmental flow principles and rules typically entails a trade-off among competing uses. Pressures are especially pronounced along rivers where water has been over-allocated, as rights of prior appropriation restrict prospects for transformative actions. Protection through prevention is vital in efforts to preserve ecosystem-sustaining flows in rivers that are yet to be harnessed by human infrastructure (Postel and Richter 2003).

Beyond such programmes, various small dams, weirs and stop-banks have been decommissioned and even removed, often relating the interests of a living river to socio-cultural considerations (e.g. East et al. 2015; O'Connor et al. 2015).

Two key factors account for the recurrent failings of small-scale restoration measures. First, they are largely cosmetic, addressing concerns for form over function. Higgs (2003) refers to such practices as boutique restoration, re-engineering rivers to a particular aesthetic rather than emphasizing the concerns for a living river. Unless restoration practices address the underlying causes of degradational tendencies, they are unlikely to be successful. Such practices generate museum pieces that mirror and mimic the past, as assertions of restoration ideals apply landscape gardening techniques to imprint notions of a lost paradise through particular sets of culturally defined aesthetics (Dufour and Piégay 2009). Often, such practices express a preference for stable, single-thread meandering channels (Kondolf 2006). Essentially, these efforts strive to ensure that the channel is in the 'right' place, minimizing disruption to other human activities on the valley floor. In many instances, the channel is reconfigured, in terms of both the alignment of the channel (its geometry and planform) and its internal make up. Sometimes boulders or large pieces of wood are incorporated as physical habitat features. Elsewhere, engineered log jams enhance the resistance and physical stability of the channel. Such practices are sometimes euphemistically referred to as the efforts of 'Boys with Toys', extending the gendered use of language associated with the imposition of command and control practices by a male-dominated engineering profession. Efforts to make rivers the same fail to respect local values (Brierley et al. 2013; Tadaki et al. 2014).

Second, short-term, band-aid solutions fail to tackle problems at the required scale. Coherent applications work at the catchment scale, working with the interconnectedness and interdependence of process relationships in river systems (Cohen 2012; Fausch et al. 2002). Large numbers of small-scale interventions, many of which do not work, not only fail to bring about improvements in environmental conditions, they also fail to inspire and engage the community in the process of river repair.

If restoration is perceived as a technical problem, expert guidance is required to design and implement measures to meet specified goals. External 'out-of-towners' are brought in as contractors to conduct this work, typically employed by consulting companies. In most instances, geo-technical engineers guide the restoration design, and river crews undertake the implementation, often at considerable expense. More sophisticated designs cost more money, often creating greater financial reward. Profit margins are even higher when similar designs can be used at multiple locations, reducing start-up costs. Such is the fake beauty of cosmetic fix and forget projects. In contrast, financial returns are greatly reduced if physical interventions are not incorporated within the restoration design. It pays, literally, to do something. Heavy machinery is a high cost investment, and with river crews on the payroll, it pays to use their skills.

A paradox lies at the heart of Medean approaches to river restoration. Contemporary scientific understandings emphasize the complexity, contingencies and non-linearities of an emergent and uncertain world in which catchment-specific understandings are required to underpin effective management practices (Brierley et al. 2013). Although some of the most egregious mistakes of historical river engineering might be excused on the grounds of limited understanding of the importance of aquatic ecosystems or controls upon their functionality, contemporary restoration practices cannot use the same excuse. Scientific understanding is no longer lacking (e.g. Wohl et al. 2005). Emerging technologies and open-source software provide readily available resources at negligible cost, thereby supporting place-based applications if data and understanding are processed and interpreted in an appropriate manner (Fryirs et al. 2019).

Such framings clash starkly with managerial imperatives for efficiency, effectiveness and replicability through simple, repeatable, quick-fix solutions that make the world manageable by anyone with a modicum of training. All too often, reaches are managed as a particular 'type' of river, emphasizing concerns for what the river looks like, but failing to address underlying causes and drivers of degradational tendencies (e.g. Simon et al. 2007). Prescriptive, readily transferrable applications apply a cook-book approach with a particular set of recipes for restoration design and application. Typically, over-simplistic, off-the-shelf solutions re-engineer the river using a standard set of applications regardless of the problem (Spink et al. 2009). Such practices create static, museum-like dioramas—faithful copies of the world as it was—a world managed by others, wherein forms, spaces and aesthetics are planned and implemented by a dominating authority. Sterile and conformist approaches to landscape design assert notions of a placeless world, marginalizing prospects for societal connection and engagement in the process of river repair (cf., Spink et al. 2010). As the call for restoration grows, citizens feel left out if their river isn't being 'managed' (see Box 3.5).

Box 3.5 A Fear of Being Left Out …
Building on research conducted for Eurobodalla Shire Council on the South Coast of New South Wales, Australia in the late 1990s, I was invited to talk at a workshop on local concerns for erosion in Tuross Estuary. Some residents wanted a breakwater to minimize erosion, notionally in the interests of maintaining some of the best fishing areas in the region. Ironically, it was the lack of a breakwater that had maintained the health of the estuary, which was one of the best functioning estuaries in the state. Although most people want to 'do the right thing', environmental values of blue-rinse greenies sometimes assert emotional connections to historical frames of reference—the world as it was—failing to appreciate the values of living, emergent systems. Erosion is an integral part of a dynamically adjusting river system. It is a good thing—in the right place, at the right time, at the right rate (Florsheim et al. 2008).

Given the rapid growth of the restoration industry, often there aren't enough experts to go round, so less qualified practitioners promise a quick fix, do their thing, then move on to the next job. As yet, there is no programme for professional accreditation. Many contestations would be encountered in framing the ground-rules for such a programme. Unsurprisingly, collective reputations of the sector as a whole are tarnished by individuals and groups that fail to act in an ethical manner.

In some instances, Medean practices conceive and conceptualize the process of river repair in the interests of profit rather than the greater good. Market-based practices recognize that because there are not enough resources to fix everything, everywhere, at the same time, investments must be prioritized to generate the best returns (outputs) from available resources (inputs). Various decision-making tools have been developed to determine high-priority areas and activities in a cost-effective manner, predicting the return on investment while recognizing that significant time lags may ensue before interventions generate substantial benefits (e.g. Pannell et al. 2013). In the process of managing trade-offs, measures taken to save some values may lead to the destruction of others, such that living with environmental degradation is the only option in some areas. Such are the inequities and injustices of a dog-eat-dog Medean world, selling out both the interests of a functioning river and the agency and well-being of society.

Such framings have made the process of river repair harder than it needs to be, instilling an ethos that someone else would fix the problem if they could, but often they may have neither the time nor the resources to do so. A Gaian ethos transforms this perspective, reframing concerns for river health as everyone's problem and responsibility. Societal engagement through hands on grassroots engagement (alternatively referred to as the flaxroots in New Zealand) supports efforts to live with all rivers in generative ways.

Mainstreaming of restoration activities has emerged as a societal norm in various parts of the world (Aronson and Alexander 2013). A tide of change towards a restoration ethos has greatly increased societal engagement and interactions with rivers, fostering commitment to the process of river repair. Successful restoration initiatives inspire efforts elsewhere, drawing people back to the river. Socio-cultural and psychological dimensions of restoration are core components of Gaian endeavours, recognizing that the process of river repair is never done—it is an ongoing commitment to respectful ways of living with the river as a living entity. Respectful relationships with the more-than-human world engender commitment to preventative practices that support the self-healing basis of healthy and resilient systems that are able to recover when subjected to stress (Piégay et al. 2019).

A Gaian lens embraces the land ethic (Leopold 1949). A sense of catchment consciousness (Newson 2010) conceives rivers as the lifeblood of the land. An ecosystem approach to environment repair applies a conservation first ethos, looking after the good bits and things that matter before they become a problem (e.g. Brierley and Fryirs 2005, 2009, 2016). The cost of prevention is cheaper than the cost for cure, and such measures are also more effective. As the cost of environmental repair is enormous, it pays to minimize damage in the first place.

Working with the river, embracing its capacity to self-heal, emphasizes the application of passive restoration techniques rather than invasive interventions, leaving the river alone as far as practicable. Whenever possible, it pays to accelerate and enhance inherent recovery mechanisms (Box 3.6). In a similar manner, space to move initiatives support ecological enhancement of rivers that have been modified by engineering practices, presenting an effective way to live with a river (e.g. Biron et al. 2014; Piégay et al. 2005). Giving the channel more room on the valley floor allows it to look after itself, sorting itself out as it responds to disturbance events.

Box 3.6 Enhancing the Self-Healing Capacity of River Systems by Working with Recovery and Applying Space to Move Programmes
Working with recovery presents an important basis to enhance river health (Fryirs et al. 2018). Sometimes improvements reflect unintentional responses to particular circumstances. The Millennium Drought had a devastating impact on socio-economic conditions in southeastern Australia. However, as pressures on the land were decreased and stocking rates were reduced, vegetation was able to recover from available seed stocks along the riparian corridor. This increased the diversity of instream habitat. Also, when the next significant floods occurred in this region, the rivers had become more resilient to change. Relative to pre-drought floods of equivalent magnitude, enhanced instream roughness delayed conveyance of flood peaks, reducing the extent of erosion and decreasing downstream impacts. Such good news stories communicate benefits of leaving a river alone, allowing the inherent capacity for self-healing to support recovery mechanisms. However, commitment to ongoing maintenance is required, as weed invasions threaten various ecosystem values of these riparian corridors.

Space to move, freedom space and erodible corridor initiatives are examples of programmes that 're-wild' or 're-nature' rivers by allowing channels to adjust—to express their voice (e.g. Habersack and Piégay 2007; Kondolf 2011, 2012; Piégay et al. 2005). Such accommodations with nature leave the river alone as far as practicable. Alluvial rivers are quite able to look after themselves, creating their own resistance elements, working out how many channels they want, what size they want them to be and their sinuosity (and hence their slope). Flows selectively rework bed materials and interact with floodplains, forming patterns of features which consume flow energy in the most efficient way. This creates and regenerates the dynamic physical habitat mosaic of a river, supporting conservation and biodiversity programmes.

In addition, space to move interventions now form an important component of catchment-scale approaches to 'living with floods'. Such programmes reduce flow rates by trapping water for longer periods of time closer to source. More gradual flow releases reduce the erosive potential of flood flows. Recent applications in Europe have reconstructed wetlands and restored floodplains to hold back floodwater, releasing it more slowly. Some space to move interventions manipulate instream habitat, others increase the width of the active channel zone, while more interventionist practices reconnect former channels and oxbows. Benefit-cost analyses demonstrate the effectiveness of these interventions, negating maintenance costs for infrastructure that keep the channel in place while providing additional buffering capacity in efforts to cope with uncertain futures (Buffin-Bélanger et al. 2015). Additional socio-economic and cultural benefits of such programmes include enhanced spiritual/emotional connections of dynamic, healthier rivers and recreational and amenity value associated with the use of buffer strips.

Embracing the role of beavers as ecosystem engineers presents an intriguing extension to such initiatives. Beaver dams literally rebuild habitats and reconstruct biophysical environments along watercourses in many parts of North America and Europe, disconnecting flow, sediment and nutrient transfer. Standing back and allowing this to happen—leaving it to beaver—presents a cheap and cheerful approach to restoration practice that works at the scale of the problem (Joe Wheaton, *personal communication*).

The process of 'doing' is an important part of approaches to river repair (Table 3.1). Rather than emphasizing restoration as an expert-driven technical task to be undertaken by others, a Gaian approach embraces societal commitment to living with rivers in generative ways. Although a hands-off approach is advocated whenever possible, hands-on engagement enhances appreciation of what restoration means and what it takes to achieve it. Riparian plantings, bush/forest regeneration programmes, predator and weed management programmes and clean-up activities help to nurture, rekindle, enhance and maintain connection and commitment to the process of repair. In a related manner, planting native species in urban gardens has greatly enhanced the diversity of birdlife and the wonders of the dawn chorus. In a similar vein, maintaining and re-establishing hedgerows in rural landscapes has engendered a suite of environmental benefits. Improvements in environmental conditions create a suite of positive trickle-down consequences in socio-economic and cultural terms.

The inclusive foundation of a Gaian world not only incorporates collective involvement in decision-making processes pertaining to planning and prioritization of activities, it also engages directly in processes of doing (the actions themselves), monitoring, learning and adapting to lessons learnt along the way (Table 3.1). Ongoing monitoring and maintenance support the process of river repair. Learning by doing and responding to lessons learnt are core components of effective practice. There are no better practitioners for such activities than the people who live along the river.

As noted by Egan et al. (2011, p. 1), restoration is a practice of hope, envisioning a better future, a practice of faith, as work is conducted in a world of uncertainty and a practice of love, in efforts to protect and enhance the lives of humans and other-than-human beings alike. Understanding the human dimensions of restoration practice, including social, economic, political and cultural dynamics is critical to this quest, as important questions of equity, influence and justice come to the fore. Much depends upon the governance arrangements that are in place. Restoration isn't real until it is owned by the communities that live along the river (Spink et al. 2010).

3.6   Engaging with Communities in the Process of River Repair: Emerging Approaches to Co-management and Co-governance

Economic growth that destroys ecological support systems is neither sustainable nor truly progress. Sandra Postel and Brian Richter (2003, p. 81)

Visionary approaches to river repair encompass much more than the prescribed activities of regulatory compliance and consents-based processes. For several decades now, integrative, participatory and adaptive approaches to river management have been widely accepted as best practice (Gregory et al. 2011; Hillman 2009; Smith et al. 2014). Principles of inclusivity and engagement underpin authentic, equitable and generative approaches to visioning (Hillman and Brierley 2005). A shared vision encompasses a common agenda and associated goals, strategies and actions. Unless society approves the aspirations and approach, and practises respect for local values and knowledge, prospects for sustained improvements to river health are limited. The regenerative benefits of inclusive participatory practices can create a snowball effect, reconstituting the way that communities connect to the river (Kondolf and Yang 2008). Once improvement in environmental conditions triggers awareness and care, a return to more degraded conditions is much less desirable, shifting the baseline of societal expectations in a positive direction (cf., Pauly 1995). Enhanced connections to rivers creates greater prospect to respect, cherish and look after them, ensuring that a healthy river is maintained.

Although highly modified rivers in urban areas typically have low potential for ecological uplift, they present considerable potential for enhanced societal connection and stewardship as part of the process of river repair. Typically, these watercourses no longer provide the services to society that underpinned the development of the settlement in the first place (Everard and Moggridge 2012). Systematic improvements to water quality, alongside local restoration and daylighting programmes have resuscitated many urban streams, making them alive again (see Box 3.7). Bringing society back to the river supports community building and shared learning through artistic endeavours, festivals and environmental education initiatives (e.g. Eden and Tunstall 2006; Petts 2007; Smith et al. 2016). Societal and cultural revival and regeneration, in turn, encourage healthier lifestyles and improve physical and mental wellbeing.

Box 3.7 Rivers at the Heart of Urban Renewal
Water features are increasingly incorporated into urban redesign and restoration programmes, viewing urban rivers and streams as focal points for socio-economic revival. Physical spaces around a river enhance societal connectivity, shaping the aesthetics, identity and feel of a city (Kondolf and Pinto 2017). Landscape architects and

planners integrate walkways, cycleways, gardens and recreational facilities alongside water features (Kondolf and Yang 2008). These facilities not only enhance aesthetic values, they create social spaces that lead people back to the river, increasing prospects for day to day interactions with rivers such as swimming, fishing and water sports. Such efforts reinvigorate the heart and soul of urban landscapes, attracting visitors and tourists to these highly visible places. Several recent winners of the Thiess International Riverprize incorporated socio-economic programmes alongside key interventions to improve river health (e.g. The Environment Agency and Thames River Restoration Trust in the United Kingdom (2010), the Buffalo-Niagara Riverkeeper Group (2016) and the San Antonio River Authority (2017) in the United States). There are many economic advantages in living with healthy rivers. Businesses thrive and real estate prices increase.

In some instances, piped watercourses have been daylighted, recreating urban streams. A major urban renewal and beautification scheme removed a concrete road and constructed a new stream along a 6 km reach of Cheonggyecheon Stream in downtown Seoul, South Korea (Kim and Jung 2018). In this integrative urban plan, completed in 2005, enhanced pedestrian access and provision of public recreation spaces linked two previously separated downtown areas. Improved traffic flow and enhanced public transport facilities improved air quality. Altered groundwater interactions improved the flow regime. Clean water and habitat improvements brought about an almost instantaneous enhancement in ecological conditions. Streamside areas now have cooler, more ameliorable temperatures than other parts of the city. Alongside aesthetic and recreational benefits, local citizens reconnected with the stream. Many visitors are now attracted to the area. Collectively, these impacts helped to revive the history and culture of the region, revitalizing the local economy.

Approaches to water-sensitive urban design link provision of water supply and service facilities to socio-cultural values of watercourses. For example, implementation of the ABC (Active, Beautiful, Clean Waters) Programme since 2006 has transformed Singapore into a City of Gardens and Water. A 3P (People, Public, Private) partnership approach links a robust, sustainable and affordable water supply to a programme of renewal for water resources, for the city and for the wider community. An ethos of 'valuing our waters' improves awareness of environmental values and encourages healthier lifestyles. Previous management practices sought to minimize flood risk by flushing problems away through pathway solutions using over-sized concrete canals and channelized rivers, along with centralized detention tanks and ponds. Now, basin-wide source solutions such as decentralized detention tanks, ponds and rain gardens, and receptor solutions such as flood barriers and a barrage system better manage stormwater runoff and reduce flood risk. Removal of pollutants improves water quality before the water is collected and used for non-potable purposes. These measures not only enhance biodiversity and the living environment, they also create new community spaces, bringing people closer to the water.

The Malacca River in Malaysia was an important trade hub in centuries past. As the condition of the river deteriorated, the city turned its back to its historical connections. Following the award of World Heritage Status in 2008, the city re-embraced the river, building new outdoor bars, cafes and restaurants adjacent to it. Such relations are far removed from the circumstances and connections for which World Heritage Status was granted. Societal relations to the river have transitioned from a key service role to support trade, to disrespect for the river because of its unhealthy condition, to embracing the river as part of urban revitalization projects.

Community engagement in the process of river repair not only fosters healthier connections between people and landscapes, it also facilitates social learning and knowledge transfer, creates employment opportunities and supports monitoring and maintenance activities. However, instrumental, technical and practical justifications for community participation, including assertions of cost savings through volunteer labour, largely leave untouched the underlying assumptions and power relations that underpin existing governance arrangements. A market-focussed approach to green politics within a capitalist system merely serves corporate interests as concerned citizens engage with what Kovel (2002) refers to as exploitative volunteerism and ecopolitics without struggle. For example, in various recycling projects, citizens provide free labour to waste management industries who are involved in the capitalization of nature. In a similar manner, a focus on individual lifestyle changes prefigures a sustainable world, one person at a time, through measures such as reducing one's personal carbon footprint, biking to work, not eating meat, recycling or not drinking bottled water. Subscribing to lifestyle politics places society exactly where the interests of corporate and political elites lie, focusing upon consumption within a competition-driven world that revolves around the quest for profit. This is not where the power of the people lies.

As different parties rarely wield equal amounts of political power, decision-making practices empower particular perspectives over others, such that patterns of participation tend to reflect power asymmetries (Molle 2009). In a Medean worldview, participation is viewed as a form of consultation, based on assumptions that we can know and manage it all, wherein leaders 'tell' practitioners what to do and how to do it. Strategies of persuasion can be applied to break down silos in the quest for more effective and generative outcomes, recognizing that healthy approaches to engagement acknowledge conflict as a necessary part of collaborative planning (Mouat et al. 2013). Disputes produce opportunities for meaningful disagreement that may, if harnessed productively, avoid unproductive or intractable outcomes. Various forms of intermediaries service the middle ground between top-down policy framings and bottom-up participatory engagement, including knowledge brokers, lobbyists, consultants, facilitators and negotiators (Gregory et al. 2011; Moss et al. 2009).

Molle (2009) refers to practices such as Ecosystem Management, Integrated Water Resources Management, Integrated Catchment Management and Adaptive Management as nirvana concepts, idealized situations in which politics are notionally pushed aside. The prospect and the spin are good, drawing people to believe that good things are happening in the right ways. Sadly, however, practices and outcomes on the ground are disappointing, as brave assertions prove to be illusory and hollow. All too often, the environment loses out, as decide-and-defend mentalities underpin such deliberations (Prince 2016). Strategic and transformative practices extend beyond the low hanging fruit of management by consensus. Inevitably, frustration ensues if planning gets in the way of action. Perpetual restructuring and loss of institutional memory do not help. In efforts to keep all practitioners on board, it pays to under-promise and over-deliver, building confidence in the process of engagement. In 30 years, it is likely that concern for river health will continue to be a prominent issue, but it is unlikely that people will remember details of the institutional reshuffling that took place in decades past.

In the deregulated world of environmental markets, emerging approaches to co-management and co-governance arrangements emphasize the sharing of rights, responsibilities and power between government and civil society. Cooperative arrangements entail interactions among a range of new actors, forums and mechanisms (Armitage et al. 2012). Multi-stakeholder processes accommodate diverse views, networks and hybrid partnerships among state and non-state actors in moves towards more collaborative, integrative, systems-based practices (Harmsworth et al. 2016). Benchmarks of "good" public governance include accountability, transparency, responsiveness, equity and inclusion, effectiveness and efficiency, following the rule of law, and participatory, consensus-oriented decision-making (Armitage et al. 2012). Notionally, co-governance and co-management arrangements give meaningful effect to the diverse values and meanings of riverscapes. In theory, citizens, elected officials, government agencies and corporate interests strive to achieve broad goals of ecosystem stewardship alongside economic development. Increasingly, such frameworks seek to reframe indigenous-colonial state power sharing arrangements, incorporating practices and processes of the original stewards of the land. For example, sacralized relations to water and rivers of various indigenous societies have opened up some important opportunities for river restoration, exemplified by dam removals based on treaty rights (e.g. Gilmer 2013; Pyle 1995).

Prospectively, a river-centric perspective that puts the interests of the river front and centre provides a disarming focal point for generative deliberations. In a Gaian world, inclusive approaches to societal engagement emphasize concern for things that matter, embracing humility in efforts to live with variability and inherent uncertainties. Thinking from the perspective of the river itself supports catchment-specific practices that respect diversity, building upon collective agreement not to harm the river while learning from it and helping it to find and express its own voice. In this lens, humans exercise their care-giving responsibility and respect for landscapes and ecosystems as an act of reciprocity. Giving back to the land presents a foundation for cultures of gratitude, sustaining the land as the land sustains us (Kimmerer 2011). Prospectively, indigenous framings of a more-than-human world can take us to a very different place in our relationship to rivers, contemporaneously conceptualizing rivers through multiple lenses in what Hikuroa et al. (2019) refer to as fluvial pluralism (see Boxes 3.8, 3.9 and 3.10).

Box 3.8 Re-conceptualizing Relations to the Waikato River, New Zealand

The Waikato-Tainui Raupatu Claims (Waikato River) Settlement Act 2010 ushered in a new era of co-management arrangements for the Waikato River. Prior to this time, developmental programmes conducted in the national interest had induced significant degradation of waterways (see Knight 2016). Such actions were orchestrated under a paradigm of exclusion, in which Māori voices were marginalized (Te Aho 2010). In response to the Act, the establishment of the Waikato River Authority as a corporate entity recognized Waikato-Tainui as co-managers and co-governors of the Waikato River. This body has a single purpose, to orchestrate the restoration and protect the health and wellbeing of the Waikato River for future generations, applying practices that address the social, cultural and economic injustices of alienating Māori from the Waikato River and its surrounding lands (Muru-Lanning 2016, p. 142). Associated framings build upon principles of mātauranga Māori (Harmsworth and Awatere 2013; Harmsworth et al. 2016).

In the new co-governance structure, Waikato-Tainui members became the guardians of the Waikato River in a shared arrangement with representatives of the Crown (Harmsworth et al. 2016). This determination was far from uncontested. In the process of negotiating various claims, the government repositioned and relabelled Māori interests as stakeholders, rather than rights-holders. Structural arrangements had significant consequences for conceptions of identity for some groups, reclassifying and relabelling particular Māori with newly defined roles and rights, with concerns for equitable proportional representation based on population and extent of tribal role (Muru-Lanning 2016). The reshaping of identities transformed the cultural landscapes of the Waikato, masking the "fluidity and political importance of other forms of Māori descent" (Muru-Lanning 2016, p. 119).

These arrangements reflect neoliberal notions of property and ownership of land and resources that are anathema to a Māori cosmology. They force Māori to relate and apply language based on possessive individualism and utilitarian notions of resources as commodities to protect their customary rights and relationships with their *taonga* (treasures) (Knight 2016; Muru-Lanning 2016). If Māori claimants wish to uphold their relations with ancestral waterways, they must acquiesce in having these relationships redefined as property interests for the purposes of the law—a kind of ontological submission (Salmond 2014). If they refuse, their claims will have no legal force. In negotiating these alternative realities (Salmond 2014), if Māori kin groups do not claim ownership of ancestral freshwater bodies and rights to fresh water are privatized, they may be left with nothing (Salmond 2017). This is a form of the tragedy (or travesty) of the commons (cf., Hardin 1968). Muru-Lanning (2016, p. 162) refers to the dilution of Māori values as a form of institutionalized pollution that implicitly transformed notions of space, place, territory, community, resources and landscape.

Box 3.9 Cultural Re-connection to Watercourses in Hawai'i
Largely as a product of their isolation, the Hawaiian Islands have many unique species, with around 90% of the plant species endemic (Herman 2016). Pressures upon native flora and fauna, especially the spread of exotic species, have had a profound impact upon biodiversity, especially plants, fish and birds. As a result, Hawai'i is sometimes referred to as the Endangered Species Capital of the World (Herman 2016).

The arrival of settlers in the late eighteenth century significantly altered the Native Hawaiian way of life. Forests and swamps were cleared and drainage networks were realigned. Impacts were especially pronounced after the 1848 Mahele when the land of the kingdom was divided and private property was created. Monocultural sugar plantations quickly blanketed the islands as a cash crop. Plantation agriculture compromised self-sufficient lifestyles. Most of the profits were taken elsewhere, with little investment in local facilities. Such activities were undertaken with little regard for traditional values, other than provision of labour, conflicting markedly with the belief system of Native Hawaiians which asserts that water could not be owned—it is a public trust resource that belongs to everyone, including future generations.

Shifts in world markets and declining fertility greatly reduced plantation profits in recent decades. Subsequent political and legal events have created opportunities for Native Hawaiians to reclaim their water rights, rekindling time honoured traditions and agricultural practices. Various legal cases have reaffirmed the legal principle of water as a public trust, restoring flows to traditional uses, often in places where streams have been dry for more than a hundred years (Kozacek 2014; Kyle 2013). Regeneration of taro crops and stream revegetation programmes, among a host of interventions, has enhanced local environments and socio-economic conditions. A period of "Hawaiian Renaissance" that started in the 1970s has brought about significant cultural and community benefits (Herman 2016). Local and Native Hawaiian communities are revitalized and reinvigorated as a united 'ohana or family. This highlights the regenerative power of an ethos of 'healthy streams—healthy communities'. Benefits include greater local-scale food security, cultural re-connection to the land through learning traditional practices from elders, embracing of Hawaiian values such as *aloha 'āina* (love the land), *malama 'āina* (take care of the land), *laulima* (work together) and *kōkua* (help without being asked), biodiversity improvements and education and rehabilitation benefits (Gregory 2014). These transformations have been achieved working under the principles of *Ho'okahe Wai Ho'oulu 'Āina* (make the water flow, make the land productive), fulfilling a traditional responsibility to care for our elder sibling, the *'āina* (land), fulfilling *kuleana* (responsibility) to ancestors (Herman 2016). Several rivers in Hawai'i now provide a focal point for environmental and societal (socio-cultural) revival through participatory approaches to reconciliation ecology.

Box 3.10 Contested Water Rights for the Barkandji Aboriginal People, Australia

The Murray Darling river system drains one-seventh of the Australian continent. Sometimes referred to as the bread basket of the country, over-allocation of water has resulted in deeply contested relations between users, states and other value-sets. The Murray-Darling Basin Commission has negotiated various compromise outcomes in the management of water, largely in efforts to address historical over-allocations to meet irrigation demands. An algal bloom that stretched over 1000 km in 1991 drew significant international attention to the river (Hammer 2011; Muir 2014). Subsequent imposition of a cap on additional water extractions required that new water demands must be met through conservation, efficiency improvements and water trading, including a major buy-back scheme. Sadly, tragic imagery of a major fish kill at Menindee Lakes along the lower Darling River in January 2019 is symptomatic of the recurrent failings of such schemes.

Indigenous rights to water have been recurrently marginalized in the midst of these troubled circumstances. Although several progressive moves have been taken to recognize Indigenous water rights within legislation, implementation on the ground has been challenging, as yet failing to legitimize custodianship of rivers in ways that respect reciprocal relationships to water in many instances (Jackson 2018). For example, the native title rights of the Barkandji Aboriginal People in New South Wales were recognized in 2015, after an 18-year legal case. However, accommodation of these common law rights within Australian water governance regimes has been extremely difficult, as various forms of misrecognition and non-recognition fail to redress historical legacies of exclusion and discrimination in access to water (Hartwig et al. 2018). As expressed by Badger Bates, Barkandji Aboriginal People, in an interview in January 2019: "What's the good of native title without water? … It's our life-blood—without water we have no name, no culture—nothing" (reported in Hannam 2019). The Barka (Darling River) is central to Barkandji culture, spirituality and teachings, reflecting interconnected cosmological and material relationships. It is home to the Ngatji (Rainbow Serpent), who created the lands and the rivers. The Barkandji are responsible for the Ngatji's health and wellbeing. Compromised river flows and degraded river health have a devastating impact upon these interwoven elements of cultural identity.

Boxes 3.8, 3.9 and 3.10 show how turbulent and emergent interactions frame socio-cultural relations to rivers, the visions that are applied and the paths that are taken to achieve particular outcomes. Although all three examples are shaped by moral and ethical obligations and commitments to 'do the right thing', contestations abound in enacting rights to self-determination. Significant steps towards attainment of river-centric aspirations build upon culturally framed applications of a land ethic. Holistic measures draw together multiple threads through revitalization initiatives that are collectively enacted, providing a focal point for inclusive commitment to enactment of a duty of care.

While the aspirations of these case studies are inherently Gaian, their implementation has been constrained by the workings and contested relations of a Medean world. Deliberations do not take place in a vacuum, and it is difficult to revoke the legacy of historical misgivings. Vested interests continue to impose an authoritarian dominion over the land, rights and others through assertions of legal precedence and prior appropriation. The disconnect from the lived realities of past relationships to the land challenges prospects to re-instigate cultural re-connections. Enormous efforts have been made to assert cultural and indigenous rights within the judicial and political systems. Inevitably, court-based procedures are expensive and exclusionary. Only certain voices are heard, typically protecting vested interests relative to the public good. All too often, deliberations with an economic focus and imperative treat participants as end users and stakeholders, deflecting attention away from the deeper, more generative prospects of cultural re-connections through 'rights'. These spaces offer genuine prospects to 'Find the Voice of the River'.

3.7   THE TRANSFORMATIVE POTENTIAL OF THE RIGHTS OF THE RIVER

After a time of decay comes the turning point. I Ching

Every now and then, the contested workings of a Medean world throw up enticing insights that support contemplations of what a Gaian world could look like and how it could work. Such transformative potential is evident in legal and political procedures and outcomes in New Zealand in which the Whanganui River was granted legal status as a living, indivisible entity, from the Mountains to the Sea (Box 3.11).

Box 3.11 Rights of the River: The Whanganui River As a Legal Entity
The granting of rights to the Whanganui River in New Zealand in 2017 emerged as an outcome of Tribunal hearings relating to breaches of the Treaty of Waitangi 1840. *Te Awa Tupua* (Whanganui River Claims Settlement) Act recognizes the Whanganui River as a natural entity with its own legal personality and rights. In the Act, the Whanganui River is described as 'an indivisible and living whole, comprising the Whanganui River from the mountains to the sea, incorporating all its physical and metaphysical elements'.

This expression of a river as having legal personhood with rights reflects a distinctively Māori perspective upon river systems. The Waitangi Tribunal (1999) accepted a claim affirming that the river was and remains the iwi's *taonga* (ancestral treasure), central to its tribal identity, way of life and wellbeing. The subsequent deed of settlement made the river, known as *Te Awa Tupua* (literally, River with Ancestral Power), a legal person who has rights, powers, duties and liabilities. Designating the river as a living being and recognizing the river as a legal entity with an independent voice placed the Whanganui River in a new set of relationships with human beings. The river's own needs and rights are given legal protection.

Although *Te Awa Tupua* Act declares the Whanganui River to be a legal person, this only roughly approximates ancestral realities. A *tupua* is not a person but a powerful being from the dark, ancestral realm; an *awa* (river) is not an individual, but a living community of fish, plants, people, ancestors and water, linked by *whakapapa* (Salmond 2017). In the deed of settlement, two people, mutually chosen by the Crown and the Whanganui tribes, are established as *Te Pou Tupua*, "the human (living) face" of the river, acting in its name (representing its voice) and in its interests (protecting its rights) and administering *Te Korotete* (literally, a storage basket for food from the river), along with a fund to support the health and wellbeing of the river.

A paradox lies at the heart of the Agreement (Charpleix 2018). The appointment of two guardians, one proposed by local kin groups and one proposed by the Crown, places the river in the same legal category as children, or adults who are incapacitated and have guardians to make decisions for them. The river's independent voice can be viewed as a form of ventriloquism (Salmond 2017). For Whanganui Māori, this marks a radical shift from ancestral conceptions, in which earth and sky, mountains and rivers are powerful beings upon whom people depend, and where river *taniwha* act as *kaitiaki* (guardians) for people, not the other way around. In the agreement signed with the Crown, human beings are in charge of the cosmos. It is yet to be seen whether making the Whanganui a legal person with its own rights is merely a modernist device that asserts property rights and resource consents, thereby maintaining the artificial separation of humans and nature that undermines conventional Māori values, or if it underpins a regime shift in which both people and the river have rights and responsibilities, aimed at returning the river to a state of *ora* (health, wellbeing).

As law is a modern human construct, creating and expressing the language of rights and duties that only humans understand, operationalizing these rights entrenches anthropocentric values (Rühs and Jones 2016). To date, legal recognition of nature's rights through the concept of personhood views nature akin to humans, and therefore, it is granted human rights. Environmental laws need to be enforceable in order to be readily implemented (Thornton and Goodman 2017). Much depends on available data/evidence, the way it is appraised/used and the influence of vested interests and political will. Inevitably, issues of ownership, private property rights and entitlements come to the fore in contestations over space and the meanings of boundaries (lines on maps). Prospectively, assertions of river rights may have significant transformative potential, allowing a river to express its own voice. Implicitly, this supports efforts to allow a river to thrive and regenerate to support its own existence, encompassing a right to flowing water, a right to convey sediment, a right to be diverse, a right to adjust, a right to evolve, a right to operate at the catchment scale and a right to be healthy (Brierley et al. 2018). Holistic, catchment-wide conceptualizations of a river as a living entity prompt regenerative relations to the more-than-human world. These relations are collective and inclusive—they apply to everyone. They are inherently adaptive and flexible, recognizing and working with the river as an evolutionary and emergent entity. They embrace uncertainty, rather than seeking to suppress it, framing human society as part of something that is much bigger, thereby regenerating and revitalizing the human spirit. Are such circumstances, relations and connections peculiar to particular instances, or are the underlying principles and premises generalizable, providing a window or a pathway to a Gaian world in which they can be rolled out to all rivers, everywhere?

3.8   Concluding Comment

[I]n a finite ecosystem there is no such place as "somewhere else". Fritjof Capra (1982, p. 238)

Notions of river health have changed recurrently throughout human history, and are likely to continue to do so into the future. In many parts of the world, significant improvements in environmental conditions and societal relations to rivers have occurred in recent decades. Practices and prospects in the process of river repair take quite different forms when conducted through a Medean (competitive) or a Gaian (cooperative) lens. Moving beyond a period in which society turned its back to the river, rivers are increasingly seen as agents of revival. However, despite the talent, commitment and achievements of many local groups, fragmented practices cannot achieve the scale and pace of change that is needed to enhance and maintain improvements to river health. River regeneration is much more than a matter for science, politics and governance to sort out—it entails facilitating and supporting communities and stakeholders to appreciate their rivers each and every day. An underlying core of respect and responsibility encompasses deep and embodied emotional and functional commitment to a duty of care.

In assessing the health of riverscapes as the canary in the cage and in appraising societal relations to Planet Earth, key choices must be made whether to act consciously through harmonious relations to rivers and each other within a Gaian ethos, or perpetuate ringing alarm bells in a Medean world characterized by degraded environments and marginalized communities. What will it take to generate and sustain a cooperative, Gaian future, addressing concerns for the prevailing imbalance of privilege, power and authority, nurturing ecosystems as they support human society?

REFERENCES

Acreman, M. (2016). Environmental flows—Basics for novices. *Wiley Interdisciplinary Reviews: Water, 3*(5), 622–628.

Armitage, D., de Loe, R., & Plummer, R. (2012). Environmental governance and its implications for conservation practice. *Conservation Letters, 5*(4), 245–255.

Aronson, J., & Alexander, S. (2013). Ecosystem restoration is now a global priority: Time to roll up our sleeves. *Restoration Ecology, 21*(3), 293–296.

Arthington, A. H. (2012). *Environmental Flows: Saving Rivers in the Third Millennium* (Vol. 4). Berkeley, CA: University of California Press.

Bernhardt, E. S., Palmer, M. A., Allan, J. D., Alexander, G., Barnas, K., Brooks, S., ... Galat, D. (2005). Synthesizing US river restoration efforts. *Science, 308*(5722), 636–637.

Biron, P. M., Buffin-Bélanger, T., Larocque, M., Choné, G., Cloutier, C. A., Ouellet, M. A., ... Eyquem, J. (2014). Freedom space for rivers: A sustainable management approach to enhance river resilience. *Environmental Management, 54*(5), 1056–1073.

Blue, B. (2018). What's wrong with healthy rivers? Promise and practice in the search for a guiding ideal of freshwater management. *Progress in Physical Geography, 42*, 462–477.

Bowler, P. J. (1992). *The Fontana History of the Environmental Sciences.* London: Fontana.

Brierley, G. J., & Fryirs, K. A. (2005). *Geomorphology and River Management: Applications of the River Styles Framework.* Chichester: John Wiley & Sons.

Brierley, G.J., & Fryirs, K. A. (2008). *River futures: an integrative scientific approach to river repair.* Island Press, Washington DC.

Brierley, G., & Fryirs, K. (2009). Don't fight the site: Three geomorphic considerations in catchment-scale river rehabilitation planning. *Environmental Management, 43*(6), 1201–1218.

Brierley, G. J., & Fryirs, K. A. (2016). The use of evolutionary trajectories to guide 'moving targets' in the management of river futures. *River Research and Applications, 32*(5), 823–835.

Brierley, G., Fryirs, K., Cullum, C., Tadaki, M., Huang, H. Q., & Blue, B. (2013). Reading the landscape: Integrating the theory and practice of geomorphology to develop place-based understandings of river systems. *Progress in Physical Geography, 37*(5), 601–621.

Brierley, G., Tadaki, M., Hikuroa, D., Blue, B., Šunde, C., Tunnicliffe, J., & Salmond, A. (2018). A geomorphic perspective on the rights of the river in Aotearoa New Zealand. *River Research and Applications.* https://doi.org/10.1002/rra.3343

Budiansky, S. (1995). *Nature's Keepers. The New Science of Nature Management.* London: Weidenfeld & Nicolson.

Buffin-Bélanger, T., Biron, P. M., Larocque, M., Demers, S., Olsen, T., Choné, G., ... Eyquem, J. (2015). Freedom space for rivers: An economically viable river management concept in a changing climate. *Geomorphology, 251,* 137–148.

Capra, F. (1982). *The Turning Point. Science, Society and the Rising Culture.* New York: Simon and Schuster (Bantam Paperback, 1983).

Charpleix, L. (2018). The Whanganui River as Te Awa Tupua: Place-based law in a legally pluralistic society. *The Geographical Journal, 184*(1), 19–30.

Cohen, A. (2012). Rescaling environmental governance: Watersheds as boundary objects at the intersection of science, neoliberalism, and participation. *Environment and Planning A, 44*(9), 2207–2224.

Collard, R. C., Dempsey, J., & Sundberg, J. (2015). A manifesto for abundant futures. *Annals of the Association of American Geographers, 105*(2), 322–330.

Collier, K. J. (2017). Measuring river restoration success: Are we missing the boat? *Aquatic Conservation: Marine and Freshwater Ecosystems, 27*(3), 572–577.

Cook, C., & Bakker, K. (2012). Water security: Debating an emerging paradigm. *Global Environmental Change, 22*(1), 94–102.

Cronon, W. (1996). The trouble with wilderness: Or, getting back to the wrong nature. *Environmental History, 1*(1), 7–28.

Davison, N. (2019, May 30). The Anthropocene epoch: Have we entered a new phase of planetary history. *The Guardian.*

Dawkins, R. (2017). *Science in the Soul.* London: Transworld.

Dempsey, J. (2016). *Enterprising Nature: Economics, Markets, and Finance in Global Biodiversity Politics.* Chichester: John Wiley & Sons.

Downs, P., & Gregory, K. (2004). *River Channel Management: Towards Sustainable Catchment Hydrosystems.* London: Routledge.

Dufour, S., & Piégay, H. (2009). From the myth of a lost paradise to targeted river restoration: Forget natural references and focus on human benefits. *River Research and Applications, 25*(5), 568–581.

Duncan, D. J. (2001). *My Story as Told by Water.* San Francisco: Sierra Book Clubs.

East, A. E., Pess, G. R., Bountry, J. A., Magirl, C. S., Ritchie, A. C., Logan, J. B., ... Liermann, M. C. (2015). Large-scale dam removal on the Elwha River, Washington, USA: River channel and floodplain geomorphic change. *Geomorphology, 228,* 765–786.

Eden, S. (2017). Environmental restoration. In D. Richardson, N. Castree, M. F. Goodchild, A. Kobayashi, W. Liu, & R. A. Marston (Eds.), *The International Encyclopedia of Geography.* https://doi.org/10.1002/9781118786352. wbieg0622.

Eden, S., & Tunstall, S. (2006). Ecological versus social restoration? How urban river restoration challenges but also fails to challenge the science–policy nexus in the United Kingdom. *Environment and Planning C: Government and Policy, 24*(5), 661–680.

Egan, D., Hjerpe, E. E., & Abrams, J. (2011). Why people matter in ecological restoration. In D. Egan, E. E. Hjerpe, & J. Abrams (Eds.), *Human Dimensions of Ecological Restoration* (pp. 1–19). Washington, DC: Island Press.

Elliot, R. (1997). *Faking Nature: The Ethics of Environmental Restoration.* London: Routledge.

Emery, S. B., Perks, M. T., & Bracken, L. J. (2013). Negotiating river restoration: The role of divergent reframing in environmental decision-making. *Geoforum, 47,* 167–177.

Everard, M., & Moggridge, H. L. (2012). Rediscovering the value of urban rivers. *Urban Ecosystems, 15*(2), 293–314.

Fausch, K. D., Torgersen, C. E., Baxter, C. V., & Li, H. W. (2002). Landscapes to riverscapes: Bridging the gap between research and conservation of stream fishes. *BioScience, 52*(6), 1–16.

Flannery, T. F. (2010). *Here on Earth. A Natural History of the Planet.* New York: Atlantic Monthly Press.

Florsheim, J. L., Mount, J. F., & Chin, A. (2008). Bank erosion as a desirable attribute of rivers. *BioScience, 58*(6), 519–529.

Fryirs, K. A., Brierley, G. J., Hancock, F., Cohen, T. J., Brooks, A. P., Reinfelds, I., … Raine, A. (2018). Tracking geomorphic recovery in process-based river management. *Land Degradation & Development, 29*(9), 3221–3244.

Fryirs, K. A., Wheaton, J., Bizzi, S., Williams, R., & Brierley, G. J. (2019). To plug-in or not to plug-in? Geomorphic analysis of rivers using the River Styles Framework in an era of big data acquisition and automation. *Wiley Interdisciplinary Reviews: Water,* e1372.

Gilmer, R. (2013). Snail darters and sacred places: Creative application of the Endangered Species Act. *Environmental Management, 52*(5), 1046–1056.

Gregory, R. (2014). *Restoring the Life of the Land: Taro Patches in Hawai'i.* Retrieved from http://ecotippingpoints.org/our-stories/indepth/usa-hawaii-taro-agriculture.html

Gregory, C., Brierley, G. J., & Le Heron, R. (2011). Governance spaces for sustainable river management. *Geography Compass, 5*(4), 182–199.

Habersack, H., & Piégay, H. (2007). River restoration in the Alps and their surroundings: Past experience and future challenges. *Developments in Earth Surface Processes, 11,* 703–735.

Haeckel, E. H. P.-A. (1866). *Generelle Morphologie der Organismen: Allgemeine Grundzüge der organischen Formen-Wissenschaft, mechanisch begründet durch die von Charles Darwin reformirte Descendenz-Theorie.* Berlin: Druck und Verlag.

Hammer, C. (2011). *The River: A Journey Through the Murray-Darling Basin.* Melbourne: Melbourne University Publishing.

Hannam, P. (2019, January 20). 'Cultural water': Indigenous water claims finally on Darling agenda. *Sydney Morning Herald.*

Haraway, D. J. (1991). *Simians, Cyborgs, and Women: The Reinvention of Nature.* London: Free Association Books.

Hardin, G. (1968). The tragedy of the commons. *Science, 162,* 1243–1248.

Harmsworth, G. R., & Awatere, S. (2013). Indigenous Māori knowledge and perspectives of ecosystems. In J. R. Dymond (Ed.), *Ecosystem Services in New Zealand: Conditions and Trends. Manaaki Whenua Press, Landcare Research, Palmerston North* (pp. 274–286). Lincoln: Manaaki Whenua Press.

Harmsworth, G., Awatere, S., & Robb, M. (2016). Indigenous Māori values and perspectives to inform freshwater management in Aotearoa-New Zealand. *Ecology and Society, 21*(4).

Harris, G. P., & Heathwaite, A. L. (2012). Why is achieving good ecological outcomes in rivers so difficult? *Freshwater Biology, 57,* 91–107.

Hartwig, L. D., Jackson, S., & Osborne, N. (2018). Recognition of Barkandji water rights in Australian settler-colonial water regimes. *Resources, 7*(1), 16.

Harvey, D. (1996). *Justice, Nature and the Geography of Difference.* Malden, MA: Blackwell.

Harvey, D. (2007). *A Brief History of Neoliberalism.* Oxford: Oxford University Press.

Herman, D. (2016, June 1). Finding lessons on culture and conservation at the end of the road in Kauai. *Smithsonian Magazine.* Retrieved from Smithsonian.com

Higgs, E. (2003). *Nature by Design.* Cambridge, MA: MIT Press.

Higgs, E. (2011). Foreword. In D. Egan, E. E. Hjerpe, & J. Abrams (Eds.), *Human Dimensions of Ecological Restoration* (pp. xvii–xxix). Washington, DC: Island Press.

Hikuroa, D., Brierley, G. J., Tadaki, M., Blue, B., & Salmond, A. (2019). Restoring socio-cultural relationships with rivers: Experiments in fluvial pluralism from Aotearoa New Zealand. In Cottet, M., Morandi, B, & Piégay, H. (Eds). Socio-Cultural Perspectives in River Restoration. Wiley (in press).

Hilderbrand, R. H., Watts, A. C., & Randle, A. M. (2005). The myths of restoration ecology. *Ecology and Society, 10*(1).

Hillman, M. (2009). Integrating knowledge: The key challenge for a new paradigm in river management. *Geography Compass, 3*(6), 1988–2010.

Hillman, M., & Brierley, G. (2005). A critical review of catchment-scale stream rehabilitation programmes. *Progress in Physical Geography, 29*(1), 50–76.

Hillman, M., & Brierley, G. J. (2008). Restoring uncertainty: translating science into management practice. In G. J. Brierley & K. A. Fryirs (Eds.), *River Futures: An Integrative Scientific Approach to River Repair* (pp. 257–272). Washington, DC: Island Press.

Howitt, R. (2018). Indigenous rights vital to survival. *Nature Sustainability, 1*(7), 339.

Jackson, S. (2018). Water and Indigenous rights: Mechanisms and pathways of recognition, representation, and redistribution. *Wiley Interdisciplinary Reviews: Water,* 5(6), e1314.

Jackson, S., & Palmer, L. R. (2015). Reconceptualizing ecosystem services: Possibilities for cultivating and valuing the ethics and practices of care. *Progress in Human Geography,* 39(2), 122–145.

Jähnig, S. C., Lorenz, A. W., Hering, D., Antons, C., Sundermann, A., Jedicke, E., & Haase, P. (2011). River restoration success: A question of perception. *Ecological Applications,* 21(6), 2007–2015.

Jowett, I. G., & Biggs, B. J. (2009). Application of the 'natural flow paradigm' in a New Zealand context. *River Research and Applications,* 25(9), 1126–1135.

Katz, E. (1996). The problem of ecological restoration. *Environmental Ethics,* 18(2), 222–224.

Kersten, J. (2013). The enjoyment of complexity: A new political anthropology for the Anthropocene? *RCC Perspectives, 3,* 39–56.

Kim, H., & Jung, Y. (2018). Is Cheonggyecheon sustainable? A systematic literature review of a stream restoration in Seoul. *South Korea. Sustainable Cities and Society, 45,* 59–69.

Kimmerer, R. (2011). Restoration and reciprocity: The contributions of traditional ecological knowledge. In D. Egan, E. E. Hjerpe, & J. Abrams (Eds.), *Human Dimensions of Ecological Restoration* (pp. 257–276). Washington, DC: Island Press.

Knight, C. (2016). *New Zealand's Rivers. An Environmental History* (p. 323). Christchurch: Canterbury University Press.

Kondolf, G. M. (2006). River restoration and meanders. *Ecology and Society, 11*(2).

Kondolf, G. M. (2011). Setting goals in river restoration: When and where can the river "heal itself"? In A. Simon, S. J. Bennett, & J. M. Castro (Eds.), *Stream Restoration in Dynamics Fluvial Systems: Scientific Approaches, Analyses and Tools* (Geophysical Monograph Series) (Vol. 194, pp. 29–43). Washington, DC: American Geophysical Union.

Kondolf, G. M. (2012). The Espace de Liberté and restoration of fluvial process: When can the river restore itself and when must we intervene? In P. J. Boon & P. J. Raven (Eds.), *River Conservation and Management* (pp. 225–242). Chichester: Wiley.

Kondolf, G. M., & Pinto, P. J. (2017). The social connectivity of urban rivers. *Geomorphology, 277,* 182–196.

Kondolf, G. M., & Yang, C. N. (2008). Planning river restoration projects: Social and cultural dimensions. In S. Darby & D. Sear (Eds.), *River Restoration: Managing the Uncertainty in Restoring Physical Habitat* (pp. 43–60). Chichester: Wiley.

Kovel, J. S. (2002). *The Enemy of Nature: The End of Capitalism or the End of the World?* London: Zed Books.

Kozacek, C. (2014, October 28). Hawaii River restorations reflect national desire to protect water for public benefit. *Circle of Blue*.

Kyle, M. (2013). The 'Four Great Waters' Case: An important expansion of Waiahole Ditch and the Public Trust Doctrine. *U. Denver Water Law Review, 17*, 1–27.

Lane, S. N. (2014). Acting, predicting and intervening in a socio-hydrological world. *Hydrology and Earth System Sciences, 18*(3), 927–952.

Lave, R. (2016). Stream restoration and the surprisingly social dynamics of science. *Wiley Interdisciplinary Reviews: Water, 3*(1), 75–81.

Lave, R. (2018). Stream mitigation banking. *Wiley Interdisciplinary Reviews: Water, 5*(3), e1279.

Lave, R., Robertson, M. M., & Doyle, M. W. (2008). Why you should pay attention to stream mitigation banking. *Ecological Restoration, 26*(4), 287–289.

Leopold, A. (1949). *A Sand County Almanac and Sketches Here and There*. Oxford: Oxford University Press.

Light, A. (2000). Ecological restoration and the culture of nature: A pragmatic perspective. In P. H. Gobster & R. B. Hull (Eds.), *Restoring Nature: Perspectives from the Social Sciences and Humanities* (pp. 49–70). Washington, DC: Island Press.

Lovelock, J. (2000). *Gaia: A New Look at Life on Earth*. Oxford: Oxford Paperbacks.

Meppem, T., & Bourke, S. (1999). Different ways of knowing: A communicative turn toward sustainability. *Ecological Economics, 30*(3), 389–404.

Molle, F. (2009). River-basin planning and management: The social life of a concept. *Geoforum, 40*(3), 484–494.

Monbiot, G. (2013). *Feral: Searching for Enchantment on the Frontiers of Rewilding*. London: Penguin.

Moss, T., Medd, W., Guy, S., & Marvin, S. (2009). Organising water: The hidden role of intermediary work. *Water Alternatives, 2*(1), 16–33.

Mouat, C., Legacy, C., & March, A. (2013). The problem is the solution: Testing agonistic theory's potential to recast intractable planning disputes. *Urban Policy and Research, 31*(2), 150–166.

Muir, C. (2014). *The Broken Promise of Agricultural Progress*. London: Routledge. Environmental Humanities Series.

Muru-Lanning, M. (2016). *Tupuna Awa. People and Politics of the Waikato River* (p. 230). Auckland: University of Auckland Press.

Nash, R. (2001). *Wilderness and the American Mind*. New Haven, CT: Yale University Press.

Newson, M. D. (2010). 'Catchment consciousness' – Will mantra, metric or mania best protect, restore and manage habitats? In *Atlantic Salmon Trust 40th Anniversary Conference*. Oxford: Blackwell.

O'Connor, J. E., Duda, J. J., & Grant, G. E. (2015). 1000 dams down and counting. *Science, 348*(6234), 496–497.

Olden, J. D., Konrad, C. P., Melis, T. S., Kennard, M. J., Freeman, M. C., Mims, M. C., ... McMullen, L. E. (2014). Are large-scale flow experiments informing the science and management of freshwater ecosystems? *Frontiers in Ecology and the Environment, 12*(3), 176–185.

Ostrom, E. (2009). A general framework for analyzing sustainability of social-ecological systems. *Science, 325*(5939), 419–422.

Palmer, M. A., Bernhardt, E. S., Allan, J. D., Lake, P. S., Alexander, G., Brooks, S., ... Galat, D. L. (2005). Standards for ecologically successful river restoration. *Journal of Applied Ecology, 42*(2), 208–217.

Palmer, M. A., Hondula, K. L., & Koch, B. J. (2014). Ecological restoration of streams and rivers: Shifting strategies and shifting goals. *Annual Review of Ecology, Evolution, and Systematics, 45*, 247–269.

Pannell, D. J., Roberts, A. M., Park, G., & Alexander, J. (2013). Improving environmental decisions: A transaction-costs story. *Ecological Economics, 88*, 244–252.

Pauly, D. (1995). Anecdotes and the shifting baseline syndrome of fisheries. *Trends in Ecology & Evolution, 10*(10), 430.

Penck, A. (1897). Potamology as a branch of physical geography. *The Geographical Journal, 10*(6), 619–623.

Petts, J. (2007). Learning about learning: Lessons from public engagement and deliberation on urban river restoration. *The Geographical Journal, 173*(4), 300–311.

Piégay, H., Chabot, A., & Le Lay, Y. F. (2019). Some comments about resilience: From cyclicity to trajectory, a shift in living and nonliving system theory. *Geomorphology*. https://doi.org/10.1016/j.geomorph.2018.09.018

Piégay, H., Darby, S. E., Mosselman, E., & Surian, N. (2005). A review of techniques available for delimiting the erodible river corridor: A sustainable approach to managing bank erosion. *River Research and Applications, 21*(7), 773–789.

Poff, N. L. (2018). Beyond the natural flow regime? Broadening the hydro-ecological foundation to meet environmental flows challenges in a nonstationary world. *Freshwater Biology, 63*(8), 1011–1021.

Poff, N. L., Allan, J. D., Bain, M. B., Karr, J. R., Prestegaard, K. L., Richter, B. D., ... Stromberg, J. C. (1997). The natural flow regime. *BioScience, 47*(11), 769–784.

Postel, S., & Richter, B. (2003). *Rivers for Life: Managing Water for People and Nature*. Washington, DC: Island Press.

Prince, R. (2016). The spaces in between: Mobile policy and the topographies and topologies of the technocracy. *Environment and Planning D: Society and Space, 34*(3), 420–437.

Purdy, J. (2015). *After Nature: A Politics for the Anthropocene.* Cambridge, MA: Harvard University Press.

Pyle, M. T. (1995). Beyond fish ladders: Dam removal as a strategy for restoring America's rivers. *Stanford Environmental Law Journal, 14*(1), 97–143.

Robertson, M. M. (2006). The nature that capital can see: Science, state, and market in the commodification of ecosystem services. *Environment and Planning D: Society and Space, 24*(3), 367–387.

Rühs, N., & Jones, A. (2016). The implementation of earth jurisprudence through substantive constitutional rights of nature. *Sustainability, 8*(2), 174.

Salmond, A. (2014). Tears of Rangi: Water, power, and people in New Zealand. *HAU: Journal of Ethnographic Theory, 4*(3), 285–309.

Salmond, A. (2017). *Tears of Rangi: Experiments Across Worlds.* Auckland: Auckland University Press.

Shapin, S. (1998). Placing the view from nowhere: Historical and sociological problems in the location of science. *Transactions of the Institute of British Geographers, 23*(1), 5–12.

Simon, A., Doyle, M., Kondolf, M., Shields, F. D., Jr., Rhoads, B., & McPhillips, M. (2007). Critical evaluation of how the Rosgen classification and associated "natural channel design" methods fail to integrate and quantify fluvial processes and channel response. *Journal of the American Water Resources Association (JAWRA), 43*(5), 1117–1131.

Slater, L. (2013). 'Wild rivers, wild ideas': Emerging political ecologies of Cape York Wild Rivers. *Environment and Planning D: Society and Space, 31*(5), 763–778.

Smith, B., Clifford, N. J., & Mant, J. (2014). The changing nature of river restoration: Changing nature of river restoration. *Wiley Interdisciplinary Reviews: Water, 1*(3), 249–261.

Smith, R. F., Hawley, R. J., Neale, M. W., Vietz, G. J., Diaz-Pascacio, E., Herrmann, J., … Utz, R. M. (2016). Urban stream renovation: Incorporating societal objectives to achieve ecological improvements. *Freshwater Science, 35*(1), 364–379.

Spink, A., Fryirs, K., & Brierley, G. (2009). The relationship between geomorphic river adjustment and management actions over the last 50 years in the upper Hunter catchment, NSW, Australia. *River Research and Applications, 25*(7), 904–928.

Spink, A., Hillman, M., Fryirs, K., Brierley, G., & Lloyd, K. (2010). Has river rehabilitation begun? Social perspectives from the Upper Hunter catchment, New South Wales, Australia. *Geoforum, 41*(3), 399–409.

Strang, V. (2004). *The Meaning of Water.* Oxford: Berg.

Tadaki, M., Brierley, G., & Cullum, C. (2014). River classification: Theory, practice, politics. *Wiley Interdisciplinary Reviews: Water, 1*(4), 349–367.

Tadaki, M., & Sinner, J. (2014). Measure, model, optimise: Understanding reductionist concepts of value in freshwater governance. *Geoforum, 51,* 140–151.

Te Aho, L. (2010). Indigenous challenges to enhance freshwater governance and management in Aotearoa New Zealand—The Waikato river settlement. *The Journal of Water Law, 20*(5), 285–292.

Thornton, J., & Goodman, M. (2017). *Client Earth.* London: Scribe.

Wohl, E., Angermeier, P. L., Bledsoe, B., Kondolf, G. M., MacDonnell, L., Merritt, D. M., ... Tarboton, D. (2005). River restoration. *Water Resources Research, 41*(10).

Worster, D. (1993). *The Wealth of Nature: Environmental History and the Ecological Imagination.* Oxford: Oxford University Press.

Wright, R. (2004). *A Short History of Progress.* Toronto, ON: House of Anansi.

CHAPTER 4

A Strategy to Express the Voice of the River

Abstract *Finding the Voice of the River* envisages revitalized societal relationships to living rivers, wherein approaches to engagement and governance embrace holistic, catchment-specific approaches to maintaining and improving river health. Using an analogy with medical practices, approaches to diagnosis, treatment and monitoring apply a duty and culture of care to attend to things that matter. Supported by uptake of rapidly emerging place-based technologies, applications of a multiple knowledges ethos endeavour to move beyond distress associated with environmental loss (solastalgia) towards assertions of sentient rivers.

Keywords River health • River management • More-than-human • Conservation • Restoration • Sentience

4.1 LIVING WITH RIVERS: A STRATEGY TO FIND AND EXPRESS THE VOICE OF THE RIVER

Our planet is a lonely speck in a great enveloping cosmic dark. In our obscurity, in all this vastness, there is no hint that help will come from elsewhere to save us from ourselves. Carl Sagan, 1996 (Pale Blue Dot, Ballantyne, New York; quoted in Dawkins 2017, p. 29)

Although enhanced environmental consciousness has engendered moves towards an era of river repair in recent decades, contemporary

© The Author(s) 2020   111
G. J. Brierley, *Finding the Voice of the River*,
https://doi.org/10.1007/978-3-030-27068-1_4

approaches to river management are unlikely to engender a sustainable future (Chap. 3). Competitively framed practices continue to exploit rivers to meet particular human interests, emphasizing socio-economic concerns through technocentric applications, negating cultural and environmental values as secondary issues. Dominion over nature remains the prevailing ethos, as command and control practices apply an alternative form of master-servant relationship to deliver ecosystem services to meet human needs.

We live in strange times—a period of remarkable potential, yet a time of palpable disquiet and unrest associated with various environmental, socio-cultural and political crises. Many people are anxious to engage in transformative practices, yet somehow do not know how to go about it. The Chinese characters that make up the term used for 'crisis', *wei-ji*, incorporate 'danger' and 'opportunity' (Capra 1982, p. 26). This is entirely apposite, as new perspectives and practices are required to cope with increasingly perilous times, characterized by profound disparity between intellectual power, scientific knowledge and technological skills on the one hand, and prevailing sources of wisdom, spirituality and ethics on the other. Present day framings, governance arrangements and practices are destroying the socio-ecological systems upon which human and environmental wellbeing depends. The endeavour that brought about the human-dominated world of the Anthropocene needs to be reframed to create a very different set of relationships.

Considerable disillusionment and disengagement characterizes increasingly anxious times (Box 4.1). Environmental crises associated with depleted resources, climate change, land use change and declining soil fertility are accompanied by growing concerns for elitism, affordability and psychological wellbeing (Harari 2018). This is an increasingly divided world—not just the haves and the have-nots, but the powerful and the powerless, the engaged and the disengaged, the connected and the marginalized, as selfish behaviours diminish priceless values of decency, dignity, loyalty, honour, morality and love.

Box 4.1 Technological Potential, Media Manipulation and the Malicious Erosion of Promise
In the interconnected and increasingly interdependent (globalized) world of the Anthropocene, technological developments have transformed our day to day lives—how we travel, work, interact. Recent decades have seen unrivalled growth in food production, greater

provision of water and energy services, enhanced capacity to address medical disasters and levels of peace and prosperity that are unprecedented in human history (Harari 2014). At the same time, migration and refugee crises have engendered increased extremism in political, religious and socio-cultural terms. Although crime rates are falling in many areas, proponents of fear loudly proclaim the imperative for greater security and increased surveillance. Despite our growing population and our capacity to feed ourselves, obesity-driven deaths and associated health problems outnumber those linked to starvation. As water and energy supply increases, so does per capita use. We have remarkable capacity to cope with disease and spread of viruses, yet water quality/quantity problems continue to cause unconscionable numbers of water-related deaths and debilitating diseases. Despite the potential for enhanced collective wellbeing, inequity is growing. Ironically, in this increasingly unequal world of power and privilege, there has been significant pushback against global movements such as the United Nations.

As evident throughout human history, science, technology and engineering can be agents of good or evil. The development of the nuclear bomb and weapons of mass destruction make us all too aware of our capacity for acts of bastardry and prospects for self-annihilation. Bacteria and viruses crafted as agents of biological warfare kill humans regardless of their age or sex, their guilt or innocence. Bombs and gas do not distinguish civilian from soldier.

The Digital Age has created information at our fingertips, presenting remarkable liberating potential for democratization of knowledge in this Age of the Masses (Harari 2014). Perhaps inevitably, powerful forces inhibit such prospects, fuelling a post-truth world of fake news, alternative realities and false gods/idols, manipulating and exploiting information delivery services and social media. Filters selectively applied by a small number of manipulators diminish the privilege of open access to information, often under the auspices of protection from hackers and security threats or as part of re-education programmes. The court of public opinion is open to abuse at the hands of the gatekeepers of knowledge. Life isn't made easy for whistleblowers. Factual accuracy is open to interpretation in an increasingly fractured and partisan media environment, as opinion pieces, commentaries and inordinate trivia dilute understandings based on hard-won evidence and critical thinking. Inconvenient facts are labelled fake news and disregarded, replaced by alternative realities. All too often, belligerent yet hollow assertions take centre stage, providing simple, convenient and appealing answers that help many people to cope with the angst and anxiety associated with the overwhelming pace of societal change. Proclamations to make things great provide a powerful rallying call for a message of hope. Essentially, this is a form of tribal signalling, wherein bias and spin serve the interests of powerful groups (Nucitelli 2017). Merchants of doubt cloud and mask inconvenient facts, diminishing prospects and diluting potential to address concerns for things that matter (Oreskes and Conway 2011). In a related manner, communicating uncertainty is difficult, as not necessarily 'knowing' the answer to a problem is sometimes construed as a threat to credibility. Although risk and uncertainty are integral parts of professional and personal lives, when it comes to environmental issues, especially legal ramifications thereof, there is an expectation of certainty in rules and outcomes. Viewing uncertainty as a creative stimulus, rather than a threat, can help in the designation of flexible, open-ended approaches to scoping prospective futures.

Unless we take control of our own agendas, algorithms that predict who we are will increasingly manipulate us in the interests of the programmers. Those who craft and write the algorithms of today design and create the workings of tomorrow (O'Neill 2016). Such technological elements of modernity echo earlier phases of human development, wherein writing was born as the maidservant of human consciousness, but increasingly became its master (Harari 2014, p. 148).

Perhaps the most alarming aspect of our times is the crippling ineptitude of existing political and governance arrangements at local, regional and (inter)national scales, and the increasing cynicism with which citizens view their leaders. From whence will come the inspiration and leadership to meet the potential that is possible?

Capra (1982) envisages a turning point for our planet, wherein a cultural revolution brings about profound changes to social and political structures. This revolution transforms our thoughts, perceptions and values from a mechanistic worldview based on rational, analytic science to a holistic conception of reality that embraces a systems view of life, mind, consciousness and evolution, appraised within a spiritual framing that is innately ecological and feminist (see Plumwood 2002). By extension, a holistic lens applies a duty of care in societal relations to rivers and each other, building on principles of coevolution, reciprocity and mutual inter-dependence. Such applications draw upon the achievements of reduction-ist science, incorporating them within an organic lens that is shaped by mystical views and understandings of all ages and traditions, embracing intuitive knowledge alongside rational thinking (Berkes 1999; Norgaard 2006). The generation of healthy waterways and the emergence of river communities revitalize the soul, shifting the baseline of societal expecta-tions to a positive, more aspirational frame of reference.

Inevitably, the process of river repair is far from a straightforward quest. Indeed, Hillman (2009) questions whether this transition is real or illu-sory. Sadly, degradational influences reflect political imposition of particu-lar sets of socio-economic and cultural values that prioritize individual concerns over the public good, such that the environment is the big loser. The longer the delays in enacting transformative practices, the more restrictive the available options, the harder they will be to achieve, and the greater the costs that will be incurred. The oft-quoted maxim of riparian vegetation planning rings loud and true: The best time to plant trees was 30 years ago! It's time to get on with it. It took several thousand years to refine human endeavours to manage rivers to meet societal needs, yet the process of river repair is still in its infancy. This chapter presents an action plan to achieve healthier rivers. Emphasis is placed upon 'living with living rivers' in authentic, generative and healthier ways. A call for socio-cultural and environmental reformation represents something akin to a 'new romantic movement' (Box 4.2). Hope springs eternal.

Box 4.2 Counter-Revolutionary Perspectives of a Technocentric Age: Prospects for a New Romantic Movement That Embraces the Values and Needs of the More-than-Human World

The Romantic Movement emerged as an intellectual and artistic movement in the second half of the eighteenth century. It reflected a socio-cultural response to multiple factors: Scientific rationaliza-tion during the Age of Reason and subsequent Enlightenment, the Industrial Age, urbanized and factory lives, loss of artisanal craftsmanship, disconnection from the land, pollution, poverty and deprivation. Mechanization, timekeeping and the division of labour objectified people and nature. Reduced to commodity status, humans were themselves degraded, spiritually alienated from the land.

The romantics sought expression in literature, music, painting and drama to convey a pure Nature as a spiritual source of renewal. Beauty was found in wild places that had long been despised as malevolent or evil. Nature, so long feared as the killer of humans and the literal place of the devil, was transformed into the perfection and uncorrupted work of God's creation, divorced from the corruptions of human endeavours (Budiansky 1995). Perceptions of wilderness were transformed from a place of chaos, ugliness and evil—unimproved wastelands, devoid of value—to one of order, harmony and beauty—places of calm, yearning and escapism. Idyllic Arcadian visions of a sublime and unspoiled wilderness reflected contemplations of a lost and virtuous utopia. These idealized conceptualizations served a range of aesthetic, spiritual, ideological and even nationalistic urges. Mountains came to be worshipped as expressions of the sublime. Poets such as William Wordsworth and Samuel Taylor Coleridge articulated the image of Nature as a harmonious system animated by spiritual forces. The traditionalism of rural life was revered, juxtaposed against monstrous machines, the dark satanic mills of William Blake and the scientifically disturbed machinations of Mary Shelley's *Frankenstein*.

Prospectively, in our increasingly technocentric times, a New Wave Romantic Movement could reconceptualize our relationship to nature and the environment more generally, reacting against an electronic and robotic age that increasingly divorces our lives from our surroundings. Conceptualizations of reality that contend that technology has freed us from dependence upon the world around us are delusional. However, the interests of progress, growth and development have blinded us to such thinking. As technology plays an ever-greater role in busier and increasingly urbanized lives, it does so at the expense of other things, like finding time to contemplate and reflect, reconnecting to notions of presence that appraise our place in the world. Paradoxically, an increasingly interconnected and interdependent device-laden 'virtual' world, with connectivity at our fingertips, has engendered increasing levels of marginalization, alienation, disempowerment and disengagement, expressed as malaise, loneliness, malcontent and melancholia—the opposite to being 'at one' with the world around us. A colleague at a staff meeting at Macquarie University in 2003 commented: "It isn't real until it's virtual and measureable."

In the art world, modernist expressions of relations to the more-than-human world in the early twentieth century highlighted the sensation and emotion conveyed by and through landscapes, wherein ideas, moods and dream states take primacy over material realities. Sensuous expressions of metaphysical domains and poetic embodiments of liberty and expansiveness emphasized concerns for living entities, even in still-life paintings, seeking to realize the sensation rather than merely capturing the object, moving beyond representations of the visible world to create more abstract, non-objective art. In this sense, paintings are conceived as alternative pathways to spiritual reality, seeing art as something that is innately sensory in all dimensions, not merely visual. Such relational connections extend beyond dystopic technocentric realities of the Digital and Electronic Age. Inevitably, challenges are faced in contemplating let alone achieving such alternative realities. At another staff meeting, this time at the University of Auckland (2009), a colleague commented: "There's no place for emotions and poetry in science." Yet, ecopoetic expressions of a more-than-human lens conjure up emotive imagery and ecological imaginaries of holistic conceptualizations that extend beyond the realm of scientific precision and epistemological certainty (Dicks 2014).

The convergence of several threads indicates considerable prospect to achieve the transformation in mindset and practice that is required:

1. The Information Age presents unprecedented opportunities to communicate best available understandings to everyone. Available information bases and remotely sensed monitoring techniques, alongside machine-learning and modelling applications, present real-time insights into the diversity, adjustments and evolutionary

trajectories of each river system. This unprecedented capacity creates the tantalizing prospect to allow each river to 'speak for itself'—to have its own voice heard!

2. Emerging social movements are aspirational and youthful. They are sick of excuses for inaction and inept practices. They no longer expect governments to 'fix' problems and they relish the chance for meaningful engagement in the process of river repair. There is significant awareness of the problems presented by ailing river health, and effective measures have been applied to demonstrate what is achievable when underlying causes of degradational tendencies are addressed.

3. Increasing recognition of the fundamental importance of a holistic ecological worldview, and appreciation of the emergent, contingent and uncertain nature of river adjustments, has drawn attention to the self-healing properties of rivers as organismic systems. Whenever possible, it pays to work with the recovery trajectory of the river, minimizing invasive interventions while helping the river to help itself. A conservation ethos protects and looks after things that matter. Significant benefits can be achieved by giving the river more space to move. Such practices free up time and resources to apply proactive measures to strategically address threatening processes elsewhere in the catchment.

As yet, a Gaian mentality is yet to emerge as a day to day reality (see Chap. 3). Rather, this prospect has been side-tracked as one of many failings of the neoliberal agenda. However, initial vestiges of this journey are already underway, with many remarkable examples of transformative practice. What is needed now is rapid expansion in the scale and scope of such endeavours, living with all rivers everywhere in an ongoing commitment to a grand socio-cultural experiment. Although written facetiously, perhaps it is time to "Make Rivers Great (Again)".

Significant transformations in practice are unlikely to emerge while working within existing structures, as powerful forces protect vested interests and resist change in efforts to maintain the status quo. As observed by Albert Einstein (1879–1955), you cannot solve a problem within the mindset that created it. Similarly, as asserted by Buckminster Fuller (1895–1983), rather than fighting the existing reality, it pays to create a new model that makes the existing model obsolete. Collective commitment to conscious living and wellbeing entails living respectfully with all rivers.

Seen through a lens of Earth Jurisprudence (Bosselmann 2011), governance arrangements encompass a different set of power relations and property rights, emphasizing concerns for the common good while respecting inherent values of the more-than-human world.

Change happens best when people want it. There is little prospect to improve river condition in sustainable ways unless society approves such aspirations. The attitudes engendered through day to day lifestyles are the key to sustained change. Ultimately, meaningful change on the ground reflects the mentality of river communities enacting ongoing commitment to a duty of care for 'their' river, scaled up through collective societal engagement for all rivers through enabling policy and legal frameworks (i.e. linking bottom-up and top-down framings). Umbrella organizations play a key role in co-ordinating such actions, whether formally or informally constituted.

Finding the Voice of the River entails taking steps to:

1. Work together, pooling expertise, dreams and visions through networks of action that upscale effectively, building upon a shared culture of kindness, care, compassion and decency.
2. Know and plan effectively through place-based endeavours that build shared understanding of things that matter, assessing what is realistically achievable and how to go about achieving it.
3. Act wisely, living with the river in mutually beneficial ways, day-by-day, applying proactive framings and precautionary principles that protect and enhance the values of the river.
4. Monitor and learn effectively, collectively responding to lessons learnt, reinforcing and adapting as required.

Prior to providing details on these steps, notions of river health are reconceptualized in relation to approaches to living with rivers rather than seeking to manage them.

4.2 Reconceptualizing the River Health Metaphor in Relation to Living with Rivers Rather than Managing Them

(H)uman and environmental health are inextricably linked. Tim Flannery (2010, p. 108)

Metaphors help to communicate complicated concepts through the use of everyday language and ideas. They play an important role in narratives and storytelling, supporting our moral compass on the one hand (Cronon 1992), while fuelling the imagination on the other (Dicks 2014). The health metaphor provides a useful basis to appraise societal relations to rivers, drawing attention to the differing values that are at play. Inherent complexities and challenges in defining river health prompt debate and discussion, helping the concept to resonate. Few people would argue against the quest to achieve, support and maintain healthy rivers (Blue 2018). The river health metaphor also plays a socio-political role, helping to reframe the narrow focus of scientific precision in relation to the richness of connotation that underpins societally important values—things that matter (Vugteveen et al. 2006). Although 'health' cannot be definitively defined, concerns for wellbeing and the process of healing focus attention on holistic appraisals of the life-supporting capacity of a system (see Box 4.3).

Box 4.3 Holistic Approaches to Health Care (After Capra 1982)

Before Descartes, most healers had addressed themselves to the interplay of body and soul, treating 'the whole patient' within the context of their social and spiritual environment. The division of mind and body engendered by Descartes (*Cogito, ergo sum*, I think, therefore I am) led physicians to concentrate on the body machine, neglecting the psychological, social and environmental aspects of illness. As conceived by Capra (1982, p. 140): "A healthy person was like a well-made clock in perfect mechanical order, a sick person like a clock whose parts were not functioning properly."

An engineering approach to health management conceives illness as a form of mechanical trouble and medical therapy as a form of technical manipulation, disconnecting relations between doctor and patient. Recurrent failures of the machine promote excess use of prescription medicines and surgical procedures, marginalizing or pushing aside the inherent healing power of the body itself. Such practices separate links between health and living habits, creating assertions and expectations that doctors can fix anything, irrespective of lifestyle values. Little wonder rivers became repositories for excess use of prescription medicines, such that they are awash with antibiotics (Wilkinson et al. 2018).

The shift in emphasis from the patient to the disease deflected and negated concerns for preventative measures such as hygiene, sanitation, nutrition and exercise. Thus emerged the various chronic and degenerative diseases of civilization such as heart disease, cancer and diabetes, closely linked to stress, diet, drug abuse, lack of exercise and environmental pollution (Capra 1982). Health and healing cannot be understood in reductionist terms, such as the absence of disease or infirmity. Rather, they entail consideration of a complex suite of interactions among physical, psychological, social and environmental aspects of the human condition. Effective approaches to health care are holistic and humanistic endeavours, integrating physical and psychological therapies in ways that reflect the inherent interdependence of mind and body.

Health is an innately subjective experience whose quality can only be known intuitively as it can never be exhaustively described or quantified. Feeling healthy reflects maintenance of integrity and a sense of balance among physical, psychological and spiritual components of an organism, and its relationships to its environment. The strongest feeling of being healthy occurs when all relationships are well balanced and integrated. Indeed, the etymology of the words 'health' and 'whole' can be traced to the Old English root word *hal* (Capra 1982, p. 234), implicitly linking health and wholeness. To be healthy, an organism has to preserve its individual autonomy, but at the same time, it has to be able to integrate itself harmoniously into larger systems. This entails being in synchrony with oneself—physically and mentally—and with the surrounding world. Illness occurs when an organism is out of synchrony.

Although general principles of health care are relevant everywhere (e.g. basic sanitation and hygiene), practices must be appropriately tailored through individual determinations, body by body, river by river. Just as no one understands pain better than the person who feels it, each river requires individual attention. Common intuitive concepts and terms used to assess human and environmental health reflect differing stages of medical practice, entailing concerns for diagnosis, treatment, monitoring and follow-up procedures (Elosegi et al. 2017). Effective medical programmes set out to analyse, minimize, remedy and avoid damage, applying precautionary and preventative practices that focus on the operation of the system as a whole, rather than as a collection of organs. The least healthy attribute of a system determines the performance of the system as a whole.

Several reservations come to mind in applying the health metaphor and the medical analogy in efforts to improve river condition. For example, while human aspirations for a healthy life focus upon physical, mental, emotional and psychological wellbeing, an ecosystem does not have a particular goal or an optimal state. Beyond this, many forms of medical practice are mechanistic and materialistic, typically applied within a command-and-control mentality, often driven by the quest for profit. Implicitly, the medical profession is symptom-driven and has a treatment focus, emphasizing the quest for solutions, typically orchestrated as targeted interventions rather than preventative practices. In contrast, healthy approaches to living with rivers emphasize concerns for preventative measures, applying curative procedures when required.

Holistic approaches to assessment of river health incorporate qualitative measures such as societal relations, psychological connections and aesthetic values alongside mechanistic measures that quantify particular environmental attributes, whether some form of biodiversity assessment, a geodiversity score, a cultural diversity index, a list (and financial tally) of ecosystem services, or some other metric (Blue and Brierley 2016). Although quantification is a useful tool, it is a poor master. What does 100% healthy mean (or 30%, or 47%)? Scientifically framed notions of river health engender a rational management ethos which postulates that if there is adequate knowledge of what values need to be maintained, and associated biophysical conditions are known, then management responses can be *designed* to meet these requirements such that conditions can be optimized to generate (or sustain) this particular set of values (Blue 2018). Such endeavours assume complete knowledge of an idealized system that can be controlled to meet a particular purpose.

As there is no absolute measure of river health, there are no clear and unambiguous guidelines as to how to assess it. Determining whether a particular attribute or trait is a good or bad thing may be inherently subjective, as it may only be a good or bad thing under particular sets of circumstances. Sometimes there may be too much of a good (or bad) thing, in the wrong places, at the wrong time (Fryirs 2015). Inevitably, measures of river health are inherently relational, in space and time. For example, health assessments may vary markedly before, during and after a flood, so it is important to appraise river health in relation to the 'expected' range of variability for the reach under consideration.

Effective approaches to analysis of river health incorporate societal values and relations to the river; concerns for things that matter. These are inherently place-based determinations (see Box 4.4). In some instances, iconic or totemic values provide holistic measures of river health. For example, the ability to swim safely in a river requires that both the physical structure of the river and quality of water are appropriate. Similarly, all components of life cycles must be met if viable populations of fish or other keystone species are to be maintained. In connecting to local values, it is important to relate socio-economic and cultural attributes to particular aesthetic and biophysical values, assessing societal and psychological interactions with the river. An inclusive approach to assessment of river health builds upon locally important meanings and values (O'Neill et al. 2008; Tadaki and Sinner 2014). This entails determining if something is missing, or isn't functioning as well as it could, whether socio-cultural connections to a river, socio-economic and recreational uses, ecological values, aesthetics or other factors. Analysis of limiting factors that impact upon river health, whether socio-cultural, biophysical or institutional constraints, underpin efforts to address system-specific concerns. This helps to assess what can be done to fix problems or live with them in different ways.

Box 4.4 Contested Place-Based Assertions of River Health
Appraisal of river health reflects particular circumstances, moods and perspectives. Consider the following examples:

- A daylighted stream with aquatic life and recreational facilities in an urban parkland has replaced a pipe or culvert. The history of stinking sewers is now a distant memory. The stream can be celebrated each and every day as a living entity.
- Geysers pump toxic chemicals into a hydrothermal stream in an active geothermal zone, such that only extreme microbes can survive. Although popular with tourists, these are extremely dangerous environments to humans, yet such systems are a significant source of life/biodiversity on Planet Earth. What is a healthy hydrothermal stream?

> - For many people, engineered rivers are indicative of progress and socio-economic success. Irrigation canals and drains provide water to agricultural communities. Navigable channels support the interests of the captains of trade and industry. The river may be silenced such that its voice is suppressed, but to local communities, provision of flow may be the key attribute of a healthy river—the thing that matters most.

Health assessment procedures report on status and trajectories against established trends and triggers, providing a reliable and relevant signal about the condition of a given system—whether its condition is improving or not. Careful selection and use of diagnostic indicators ensures that procedures identify underlying causes of degradation, guiding what can be done to address them (Fryirs 2015). This provides a meaningful basis to compare like with like. Early warning indicators support strategic and proactive actions, helping to identify threshold conditions and tipping points, wherein a relatively small change in a driver may induce rapid and dramatic change in the outcomes of threatening processes. As conditions change, it may be necessary to modify the range of measures that are monitored (Brierley et al. 2010).

Appraisal of river health entails concerns for product and process. A river-centric sense of place looks after things that matter. Societal engagement through participatory practices engenders collective commitment to a duty of care, living responsibly and respectfully with each and every river, acting as stewards or guardians in efforts to regenerate waterways and communities alike. Playing with the health analogy, assertions of river health can be infectious, always remembering that what is considered healthy today may not be considered healthy in the future (see Chap. 2).

4.3   Engendering Societal Engagement to Live with the River in More Generative Ways

If competition is evolution's motive force, then the cooperative world is its legacy. And legacies are important, for they can endure long after the force that created them ceases to be. Tim Flannery (2010, p. 31)

Finding the Voice of the River is not something to be determined and implemented by governments and corporations. Rather, it entails societal commitment to living with the river in respectful and responsible ways. Collective engagement through participatory practices requires effective governance structures that support linked-up thinking and practices, facilitating collaborative efforts between individuals and informal/formal institutions, engendering communities and networks of connection without getting in the way (see Bodin and Crona 2009).

Generative practices embrace concerns for procedural fairness, ensuring that individuals have a voice, and distributive fairness, such that individuals are satisfied that a fair outcome has been achieved. Such endeavours move beyond fragmented and authoritarian approaches to public participatory practices, construed as information supply and consultation that reports the findings of others, wherein managers and decision-makers act on behalf of publics, speaking for them (Huitema et al. 2009). *Finding the Voice of the River* envisages situations in which the public is in charge of determining and enacting socially and environmentally just practices, bringing their information, insight and creativity to bear in a process of learning by doing (Huitema et al. 2009). In such framings, communities have power *with* another rather than *over* another (Gaventa 2006).

When things are working well and good things are happening, rules of engagement go largely un-noticed. Once a commitment to particular habits of mind has been attained, it is hard to go back to past behaviours. For example, as littering, recycling and composting become a way of life, it becomes unconscionable, almost inconceivable, to throw things away or leave clean ups to others. As in any relationship, what happens during good times (when ideas and potentialities are vibrant, and resources are plentiful) may be less important than collective responses to challenging situations (often periods of contestation and conflict, when key decisions over priorities must be made or support is not forthcoming). Putting the interests of the river front and centre can provide a disarming platform for such endeavours, potentially defusing personal and/or institutional biases through agreement that no actions will harm the river.

Local community champions play critical roles in these deliberations, co-ordinating actions as guardians or stewards of the river. Processes of facilitation give careful attention to sharing and listening, not construing silence as consent and not speaking for others, always leaving the door ajar to accommodate more inclusive practices and outcomes (Hillman 2005, 2006). However, a commitment to democratic processes and outcomes is not necessarily sacrosanct in such deliberations. As the functionality of a river system is constrained by its most poorly performing attribute, a lowest common denominator, "one out, all out" approach provides a sound approach to assessment of a 'good condition' river. All too often, compromise solutions represent no-win outcomes in environmental terms.

It pays to work together to protect and enhance things that matter. This entails finding points of connection and building on them in the process of engagement, sharing perspectives and learning from each other. This process can be generative in its own right. Place-based concerns for river health provide a platform to appraise societal values relating notions of belonging and identity to concerns for cultural, socio-economic and psychological wellbeing (see Box 4.5). The river is not a passive player in these deliberations. It has agency, shaping societal relations, while adjusting to them. As noted by Tim Palmer (1994, pp. 221–222), "… it is *love* of place that is needed, based on revealing knowledge that can be unearthed in the story of the river." The process of revealing a river's story can be insightful in its own right, relating local values to regional- and national-scale agendas (e.g. Comby et al. 2019). Scoping prospective futures entails nuanced interactions—with rivers and with others, formal and informal, purposeful and playful.

Box 4.5 Place As a Sticky Universal (Tsing 2004)

Being-together-in-place conveys a web of ecological, cultural and bio-geophysical relationships that make 'place' possible (Larsen and Johnson 2017). Notions of place are distinct yet omnipresent, cherished yet contested, authentic yet imaginary (Morin 2009; Wylie 2007). A place is defined by societal relationships to it, framing notions of utility and functionality alongside emotional and psychological connections and aesthetics. Identifying with place is an integral part of human identity and wellbeing, related to notions of belonging and home. The term topophilia refers to the affective bond between people and place—literally, love of place (Tuan 1974). Topophilia is inherently tied to values, preferences and aesthetics. Topophilic expressions of love of place tied to affective relationships to particular environments can be contrasted with antipathies to place expressed through topophobic relationships—apprehension, dislike, revulsion or even fear of places. Presumably, such sentiments are as old as humanity.

Landscapes provide the material grounding of human history (White 1990), presenting a platform of connection in appraising relationships to place that bridge biophysical and socio-cultural connections (Wilcock et al. 2013). Human activities are constrained by the biophysical environment, but at the same time, they also shape it. Landscapes are infused with cultural meanings, selectively imbuing place-based values and connections (Morin 2009; Nassauer 1995; Schama 1995; Wylie 2007). Such meanings are inherently sensory and experiential, reflecting ways in which human activities socialize a landscape, sometimes asserting authority and ownership (Mitchell 2003; Tuan 1977; Wylie 2007). Landscapes carry symbolic or ideological meanings that reflect back and help to produce social practices, lived relationships and social identities. In terms of social life, landscapes not only 'are', they 'do', carrying with them sets of representational practices (Morin 2009).

Relationships to place are deeply situated and contextual. They are far from static, bounded entities. Many values are historically imprinted and embedded, while others are distinctively modern, reflecting recurrent renegotiations in ways of inhabiting the land. Points of connection help to define 'things that matter', helping to identify and prioritize values to protect and/or enhance in the process of river repair.

Concerns for place attachment shape notions of nostalgia. Conversely, disconnection may engender feelings of solastalgia—emotional and psychological distress associated with environmental loss when elements of a place are missing and something doesn't feel right (Albrecht et al. 2007). Nurturing a sense of connection and belonging supports spiritual needs that are unlikely to be engendered in the anonymous spaces and exchangeable environments of a placeless world (Relph 1976).

4.4   Knowing and Planning in the Development of Proactive and Precautionary Approaches to River Health Assessment: The Visioning Process and Diagnosis of Problems

In theory, there is no difference between theory and practice. But in practice, there is. Yogi Berra (1925–2015)

A revolution in practice is transforming the knowledge bases that underpin analyses of river systems. There are two key components to this reframing. First, automated remote sensing and measurement techniques support the development of catchment-specific understandings of biophysical relationships and interactions in river systems (Fryirs et al. 2019). Such datasets create the capacity for each river system to speak for itself, to tell its own story (Brierley et al. 2013; see Box 4.6). For example, a monitoring system termed the sentience project was conceptualized as the 'brain' for the Sevier River network in Utah, United States.

Box 4.6 Theory Is Good, but Reality Is Better

I was on the 13th hole of my local golf course on Waiheke Island when a thought crossed my mind. In theoretical terms, if I hit the ball correctly, it will go where I intend it to go. How does this principle translate to my professional life? Fundamentally, this was a pretty darned frustrating question, as one of the reasons that I play golf is to get away from work (much though I love rivers, it's nice to contemplate other things some of the time). But rivers permeate my consciousness, sometimes impacting upon the effectiveness of my golfing exploits (the list of excuses is *very* long). In turn, my golfing endeavours gave me cause to reflect upon notions of a 'theoretical river'. What does it look like? How does it behave? Whose values does it satisfy? How well does theory relate to the 'real world'?

As an environmentalist of the watermelon ilk, I sometimes wonder whether I should be playing what often proves to be an incredibly frustrating game. But it's definitely not the rivers that are frustrating—it's the ways we choose to live with them. Yet in some instances, theoretical framings present a satisfactory basis to account for local reality. And occasionally, my golf ball goes straight, ending up close to where I intended it to go. Practice may make perfect, but not yet, for sure.

Second, an emerging multiple knowledges ethos relates scientific and technical insights to local knowledges, supporting the development of catchment-specific models of how each river system works. A landscape template provides an integrating platform to analyse the qualities of the mutually nurturing relationships between people and their environments, appraising historical connections and layers of histories laminated on the land to appraise river evolution (Salmond 2017). Such analyses start with the earth itself (geomorphology, earth history) and associated plants and animals (ecosystems). This is followed by assessment of the layer upon layer responses of the river to human impacts, reflecting differing phases of societal relations, values and uses and the legacies that are visible today. Such framings underpin the generation of authentic visions that reflect core values for each river system, working out what is required to protect and enhance efforts to allow each river to express its own voice.

Implicitly, rivers and landscapes convey a web of stories and journeys, told and retold in many voices (after Mahood 2016, p. 193). From this emerges a cognitive framework of layering that captures and conveys the country as embodied knowledge, bound by time yet mutable at the same time. Stories morph with each retelling, in the space between real and mythic time: "In the slippage between fact and memory, a different truth emerges" (Mahood 2016, p. 138). A multiple knowledges ethos entails listening to country, wherein place-based relational understandings encompass humans, more-than-humans and all other tangible and non-tangible elements which are co-becoming (Country et al. 2016). Approaches to cooperative learning, collaboration and reflection apply discursive processes to co-create knowledges-in-action, acknowledging different modes of knowing, learning and communication (Bishop and Glynn 2000). Multiple knowledges build upon multiple sources of information, whether qualitative or quantitative, empirical or theoretical, real or virtual. Stories, narratives, folklore, parables and proverbs can support these interactions. This requires sharing and listening with humility, respect and openness to change (Johnson et al. 2016). Participatory mapping exercises provide a basis to share understandings and assess patterns and linkages in landscapes, helping to define scales of enquiry and entities of interest (Mahood 2016). Such deliberations help to develop a shared sense of factors that have influenced river condition and measures that can be taken to address particular problems.

Key choices must be made in relating formal knowledge structures that underpin the rational workings of rigorous, quantitative scientific understandings to emotional connections and sensory feelings experienced through interactions with particular places. Much learning is innately experiential, reflecting instinctive and intuitive understandings (it's in the blood, you know). Feelings, emotions and psychological assertions sit alongside rational determinations. It is intrinsically difficult to rationalize goosebump moments captured by expressions such as: "Someone walked over my grandmother's grave", "This place gives me the creeps", "I have chills or shivers or a tingling down my spine", or "My hair is standing on end". Failure to act upon such intuitive understandings, however irrational they may seem, can result in immense frustration, personal recriminations and self-doubt.

Co-produced knowledges bring together insights from a multitude of sources, recognizing explicitly that compromise perspectives and outcomes based upon synthesizing, combining or integrating knowledge systems are not always acceptable and tenable (Box 4.7). As noted by Mahood (2016, p. 216): "In the gap between two ways of seeing, the risk is that you see nothing clearly."

Box 4.7 Professional Frustrations at the Interface of River Science and Management

The post-normal dilemma is alive and well, with much 'applied' research conducted independently of managers and practitioners (Rogers 2006). I closed the final session for the EU REFORM Project Workship in Wageningen (2015) by stating: "If struggling to translate, it's too late".

Within science itself, sub-disciplinary boundaries, knowledge structures, the arrogant assertiveness of discipline-bound practitioners and the institutional framings and constraints that go along with such dilemmas have presented some of the most significant frustrations of my career. In my experience, the holistic ecological worldview that Capra (1982) asserted was imminent has proven to be remarkably illusory, and it has proven to be extraordinarily difficult to engender constructive interactions at the interface between geomorphology, hydrology, engineering, ecology and water quality specialists. Such deliberations come down to people, with generative encounters sometimes reflecting outwardly 'odd' interactions. An interesting and productive project that worked across the arts-science divide entailed choreographed expressions of Māori relations to rivers conveyed through dance as part of an Arts Festival Event in Auckland (Longley et al. 2013).

In order to look after a river, protecting things that matter and enhancing river health, it is important to know the underlying causes of problems in order to treat them effectively: what is missing, what isn't working well, what caused it, why there, if it was left alone would it fix itself, over what timeframe? Catchment-scale conceptual models apply place-based, system-specific understandings to generate appropriately tailored responses (see Mika et al. 2010). Foresighting exercises can be used to predict future scenarios and assess possible risks, threats and responses: if we do this here, what off-site impacts are likely, where, over what timeframe? Such interrogative exercises can help in the design and implementation of Catchment Action Plans, striving to ensure that actions taken to fix one problem have minimal undesirable side-effects, such as exacerbating other problems, creating new ones or counteracting treatments for other ailments (Palmer et al. 2005; Schmidt et al. 1998). Iterative testing and refinement of the model supports ongoing learning. Careful commitment to experimental procedures learns effectively from experiences. Trials and controlled case studies consider how readily and reliably findings and implications can be transferred from one situation be another.

Planning underpins proactive and precautionary approaches to look after things that matter. Although many lessons can be learnt from the past, extrapolations should be approached with caution, remembering that decisions made and practices applied are usually undertaken with good intent, but many visions from the past may not appear to be quite so visionary today (Walker et al. 2015). Flexibility is required in scoping prospective futures, recognizing that it is disingenuous to lock-in aspirations, expectations and goals as circumstances will inevitably change (Brierley and Fryirs 2016). It pays to remember that the human gene pool reflects the survival of ancestors who adapted and evolved to conditions in the past, presenting what Dawkins (2017, p. 82) refers to as the 'Genetic Book of the Dead'. Such a genetic predisposition does not necessarily provide appropriate guidance for a future in which conditions will be different, in ways that are not entirely knowable.

Inclusive and generative planning processes engender shared responsibility for actions and outcomes, shaping efforts to move forward together with a common sense of purpose. Co-creation of a shared and strategic vision builds upon collective understanding of what is valued for each river system, alongside understandings of evolutionary traits and emergent properties. Such framings work with variability and uncertainty rather than strive to suppress them.

Rather than being an endpoint, a plan is a working document, forever in development. After all, perfection is the enemy of the good (attributed to Voltaire). Creating space for flexibility allows the capacity to monitor outcomes and tailor treatments accordingly, responding to strategic opportunities as they arise. All too often, proactive plans are perceived as good wisdom until difficult circumstances arise, yet these are the very instances for which such plans have been developed (Box 4.8). At these times, it is important to hold steady in efforts to address concerns for agreed-upon goals, unless there is a good reason to do otherwise. Once attributes that matter are lost, it may be difficult or impossible to regain them. Sound protocols are required to cope with emergency situations, as over-reactive, short term 'solutions' typically apply measures that do not work in the long term, often making things worse. Throughout these endeavours, it is imperative to stop doing stupid things, such as clearing native vegetation, over-extracting water or allowing incursions of exotic plants and animals.

Box 4.8 Communicating Proactive River Management

I used my Opening Address at the Pacific NW River Restoration Conference in 2013 to convey a recent experience from Yunnan in western China in which a colleague was scheduled to make a documentary about a successful river rehabilitation initiative that had reinstigated a step-pool sequence along the course of a former debris torrent (reported in Yu et al. 2012). Unfortunately, the local authorities hadn't realized the enormous benefits that had been achieved by allowing keystone boulders to renaturalize the rivers, and the significant geo-ecological recovery that had ensued. Rather, in efforts to 'protect' what had now become an attractive river with fish, the channel was dismantled and stabilized within riprap to make a recreational park. The audience liked the story, chuckling away at the failings of reactive management and the importance of communication. However, the response was much more muted when my next story recounted a similar encounter along Logan Creek in Utah, in which reactive management activities following a flood event removed remaining riparian vegetation and instream wood that had accumulated over thousands of years. Ironically, the Center for Watershed Resources at Utah State University lies within this watershed.

In relation to a medical analogy, doctors carefully weigh the risks and benefits of taking action, as an incorrect diagnosis leads to an incorrect prescription and can cause harm. Contextual circumstances are critical in weighing risks. When the potential benefits are large and alternatives are lacking, measures involving high risk may be acceptable. Application of triage principles may help to prioritize targeted interventions, remembering that all patients/rivers are important and there are obvious ethical implications in moving resources away from patients/rivers with low or limited recovery potential. Physicians are legally obliged to inform their patients about both the benefits and risks of a proposed treatment and any alternatives. In a similar vein, wide communication of benefits and risks of a proposed intervention is an important part of river restoration practice.

Diagnosis entails identification of problems and determination of the most effective approach to treat them. This is the first step in curing a disease or treating an ailment. Medical diagnosis starts with anamnesis, where physicians examine the medical history of patients and inquire about personal matters such as general constitution, profession and lifestyle before asking questions about the particular health problem (Elosegi et al. 2017). This helps to contextualize current symptoms and assess health risks. Differential diagnosis procedures appraise multiple causative factors. Essentially, three questions must be asked before an appropriate treatment can be determined: what is the problem, why is it a problem and how did the problem arise (what were the drivers or controls upon that situation)? Some ailments may be immediately evident (e.g. broken bones), but they may have arisen for different reasons (e.g. an accident, or osteoporosis), reflecting different drivers or controls (e.g. poor road conditions, fog or bad driving; old age or coeliac disease). Similar complexities are encountered in assessing river health. For example, an algal bloom may be obvious or self-evident, but it is important to know whether it is a response to a particular circumstance (e.g. a one-off biogeochemical situation) or whether it occurs recurrently at that location (i.e. it is expected at particular times or under certain conditions). Many symptoms can have multiple causes in response to multiple stressors. For example, an algal bloom may be a product of climate/flow variability, a response to contemporary agricultural practices, or reworking of nutrients stored along a river course as a result of past agricultural activities (Elosegi et al. 2017).

Co-ordinated planning procedures work out where to start and the sequence of activities that follow, with an appropriate rationale and evidence base. A conservation first ethos looks after the good bits before they become a problem: prevention is cheaper, and more effective than cure. This builds directly upon knowledge of values to be protected and enhanced in any given river system. Proactive practices strategically address pressures and threatening processes which may have a negative impact upon river health before the problem becomes insurmountable. Such efforts maximize prospects for rivers to become self-sustaining, helping the river to help itself (Fryirs et al. 2018; see Box 3.6).

In some instances, it may pay to work with the visionary and the willing to demonstrate what can be achieved, working in visible areas that enhance societal connections to the river. Whenever possible, catalytic actions engender positive outcomes at one place with trickle-down consequences elsewhere. For example, riparian vegetation and weed management programmes in one part of a catchment may help to address flow and sediment management problems elsewhere. Such determinations are an integral part of ongoing commitment to the healing process.

4.5   The Healing Process: Treating River Health to Protect and Enhance the Voice of the River

After reaching an accurate diagnosis, choices must be made in deliberating among treatment options. Some types of treatment are curative in the sense that they solve a given problem, others are palliative, meaning that damage or symptoms are only alleviated. Much can be achieved in improving the aesthetics of a riverscape, making it more attractive for a range of socio-cultural uses, recognizing that there may be limited prospect to improve river health in ecological terms. Although many treatments seek to address immediate issues of concern, similar to forms of pain management, a good, healthy life is the long-term goal.

In *Finding the Voice of the River*, a river-centric lens sees healthy environments and healthy communities as one and the same thing. An ethos of minimizing harm looks after and protects healthy relationships, repairing those that are unhealthy. First and foremost, this is conceived in preventative rather than curative terms. In some instances, it may be necessary to learn to live with problems, as little can be done to redress them in substantive ways. In other instances, it is important to realize that it is not necessary to intervene all the time; indeed, it may be harmful to do so. Surgery is conceived as the last option, only resorting to such measures when there are no viable alternatives. Sadly, many river restoration initiatives are forms of high-cost plastic surgery (Chap. 3). Cosmetic measures may look good for a while, but they are skin deep. Superficial practices fail to address underlying concerns for the processes that maintain a living, well-functioning river.

Rivers in a healthier state have greater capacity to self-heal. However, if elements of functionality are missing, prospects for self-healing may be limited or difficult and costly to address. Applications of the precautionary principle look after the good bits of catchments before they become a problem. While all parts of the catchment are important, high value reaches must be identified and protected. In a similar vein, vaccination programmes are important proactive and precautionary steps within comprehensive healthcare initiatives, preventing the spread of disease. Box 4.9 presents a playful commentary on the use of healthy diets and exercise regimes as a basis to create and maintain healthy rivers.

Box 4.9 Healthy Diets and Exercise Regimes for Rivers: Some Playful Contemplations with Joe Wheaton, Utah State University
Healthy bodies eat well and exercise regularly. Once we are wedded to such activities, adrenalin kicks in, fostering ongoing commitment to healthy lifestyles.

A healthy diet provides an appropriate level of nutrition that supports a self-sustaining metabolism, enabling a system to cope in times of stress. While empty carbohydrates may taste good and provide instant gratification, excess consumption through nutrient overload engenders secondary health problems that are often very expensive to address.

Rivers with the capacity to self-heal are able to provide their own healthy meals (e.g. floods that deliver water, sediment, organic matter, wood, etc.). To prepare its own meals, a river needs adequate drinks (floods) and sufficient space (i.e. floodplain areas) to generate and process its own food (i.e. sediments and wood). Healthy rivers are fed by their catchments to provide their own meals. Adjusting channels exercise regularly, with flows that interact with floodplains.

Different types of river have differing dietary needs. Just as roughage is an integral part of our diet, roughness elements exert a critical control upon energy use by rivers. Unfortunately, many rivers have a polluted and unhealthy diet, as much of their food is so fortified with preservatives (e.g. rip-rap) that it is unhealthy. Along urban streams, for example, ingredients have been depleted, removed or have gone bad, so rivers have lost their self-sustaining capacity and ingredients must be provided artificially and recurrently. In a similar vein, exercise regimes are restricted as hard structures impose stability and arrest processes of channel adjustment.

Extending the culinary analogy, the poor health of many river systems has stimulated a tremendous appetite for process-based restoration measures. Practitioners are hungry. However, the cookbook of available treatments only has a few recipes, with a narrow range of 'fixes' to a wide array of problems. All too often, prescriptive practices apply over-simplistic measures to manage a river to a particular 'type', failing to give appropriate regard to individual circumstances. A wider range of recipes could provide more effective practices across a wider range of sites. More palatable and affordable (cheap and cheerful) recipes would help to meet the dietary needs of rivers, feeding and nursing them back to recovery. Measures need to be cheap so that they address the magnitude of the degradation, and they need to be cheerful in that they need to work. Many recipes are so expensive that they can only be implemented over very short lengths of rivers, typically as sparse, boutique restoration initiatives.

In preparing a meal, much depends upon the local circumstances, available ingredients and facilities, and the chef. Although cooking techniques are relatively similar across the world (e.g. baking, barbequing, grilling, etc.), and ingredient lists have similar components (e.g. meat, vegetables, grains, fruit, herbs, spices, etc.), differing environmental conditions and socio-cultural traditions result in differing culinary preferences, practices and outcomes. In rivers, ingredients can be thought of as water, sediment, vegetation, wood, nutrients and so on, where combinations of these components generate particular features of the river. Like food production, seasonal growth patterns influence system functionality, reflecting the flow and sediment regime, variability in riparian vegetation associations, wood input and habitat availability/uptake. In many situations, the ingredients, resources, tools and time we would like are not available, constraining what is achievable in any given instance. While some ingredients can be substituted, others are so critical that the dish fails without them. The process and order in which ingredients are assembled are critical to the palatability (success) of the product. With the notable exception of a smoothie or a puree, few dishes turn out well when all the ingredients are simply blended together.

All too often, approaches to restoration emphasize quick-fix responses, motivated by profit (e.g. diet pills or invasive solutions like lipo-suction or cosmetic surgery). Applications of preventative measures do not generate as much financial return as curative practices that emphasize prescription drugs and intensive management treatments. In a similar manner, the river restoration industry profits from applications of interventionist treatments. Such actions can be considered as analogous to invasive surgery (e.g. channel realignment) or taking prescription medicine (e.g. artificial supplements like rip-rap). Without doubt, surgery is required to resuscitate some rivers, but in many instances, a treatment focus emphasizes surgery as the only solution. Often, repeat surgery is required. In the medical world, organ transplants are seldom provided to patients who are near death.

Also, surgery is something that only skilled surgeons are licensed to perform (i.e. accredited engineering contractors for rivers). However, following recipes can help to transfer complicated and nuanced understandings of river systems to inform effective actions. While the best chefs don't use a cookbook, relying on instinct, feel and flair, most of us do, or probably should. Even when various individuals follow the same procedures, the outcomes that ensue are often quite different! There are many more ways to get a dish wrong than right. Appropriate skillsets and experience are required to adapt general principles to local circumstances. Training schools (professional short courses) may help to share experiences.

Whenever possible, a living river ethos applies passive restoration techniques, consciously leaving the river alone, allowing it to generate its own healthy diet and exercise regime through freedom space or space to move initiatives (the erodible corridor concept).

In order to avoid costs of repair, strategic actions may be required to address threatening processes before problems become insurmountable. Examples may include initiatives to reduce nutrient inputs, flood management programmes such as wetland (re)creation and minimizing the introduction and/or spread of invasive species. Such efforts increase buffering capacity, providing a form of insurance in attempts to cope with uncertain futures. Much depends upon the trajectory of adjustment of the system, especially whether recovery mechanisms are improving river condition (Brierley and Fryirs 2016; Fryirs and Brierley 2016). In these instances, the conscious decision to leave the river alone may be the best option, allowing the river to sort itself out—to look after itself (Box 3.6). In other instances, it may be wise to enhance or accelerate prospects for recovery, applying measures such as fencing-off the river or weed maintenance programmes.

Equitable access to appropriate healthcare services and facilities surely helps in the process of healing and repair. If relationships are damaged and the river is unhealthy, recourse must be taken to active restoration techniques to initiate and/or accelerate recovery. Examples include riparian plantings, fish stocking, emplacement of habitat features such as large wood structures and boulder additions or manipulation of the flow regime. Loss of critical functionality may require surgery. Some treatments may entail physical removal to eliminate a problem, whether a tumour within a body or a malfunctioning (or inappropriate) dam, weir, levee/stopbank or pipe along a river. Sometimes, simply removing the causative agent such as a pathogen or environmental stressor can facilitate recovery. For example, removal of livestock allows riparian forests to regenerate, or treatment of sewage before it is discharged into a receiving stream.

To be effective, and engender sustainable outcomes, treatments must work at the scale of the problem. Lots of actions in the wrong places will not create healthy rivers. Some treatments need to be applied just once to be effective (e.g. dam removal), whereas others are prescribed forever (e.g. perpetual stocking of juvenile fish if self-sustaining populations cannot be established). Elsewhere, regenerative medicine will suffice. Sometimes, over-application of a particular treatment may reduce resistance or immunity, closing off future treatment options. New techniques may transform approaches to practice. In medical terms, keyhole surgery has greatly reduced the spread of infection on the one hand, and greatly reduced the time for recovery on the other. One thing is self-evidently clear: blanket applications of over-simplistic, off-the-shelf measures that do not to address the underlying causes of degradational tendencies are unlikely to succeed (Spink et al. 2009). Another medical analogy comes to mind: sometimes bad practices can spread like a virus or disease, and it is often very difficult to unpick their consequences.

Context-specific circumstances determine what is realistically achievable in the process of river repair. Sometimes social, economic or financial constraints limit prospects and approaches to implementation, despite readily available therapies. Much depends upon regulatory frameworks, governance arrangements, the resource base and affordability of service provision. Carefully crafted legal and governance frameworks co-ordinate networks to support such endeavours, linking bottom-up engagement to top-down policies. In Europe, policy developments within the Water Framework Directive emphasize equitable social rights for access to good condition waterbodies. Elsewhere, generative framings can build upon examples presented by policies and regulations that support anti-litter and anti-smoking campaigns, providing a nudge to reinforce the adoption of healthy lifestyle choices as issues of personal responsibility.

Restoration can be conceived as performance, wherein experiences (re) build communities through reconnections to rivers (Egan et al. 2011). As ecosystems and social systems are inextricably intertwined, restoring one helps restore the other (Bliss and Fischer 2011). Engaging productively and learning from participatory approaches to community-based river restoration and clean-up campaigns have a range of psychological and therapeutic benefits alongside social learning and environmental educational benefits. In socio-cultural terms, it is important to build, enhance and maintain socio-cultural connections to the river; indeed, this is a critical part of the rehabilitation process (Mould et al. 2018). Success helps in these endeavours, enhancing ongoing commitment to a duty and culture of care. Simply being around a living river, and making it a part of day to day lives, supports and enhances connections to place and psychological wellbeing. In turn, drawing people back to the river enhances economic prospects.

Upscaling local applications and creative practices to enhance collective impacts is challenging. This requires working collaboratively for mutual benefit, navigating the networks and building new partnerships. Sharing experiences enhances prospects to roll out lessons from successful interventions, learning from exemplars of best practice. Careful use of demonstration sites can inform activities elsewhere. This requires documentation of what activities work well under what sets of circumstances and the use of training and monitoring programmes to guide appropriate uptake of measures borrowed from elsewhere.

4.6   A Strategy to Enact the Voice of the River: Ongoing Monitoring and Learning

Unless you monitor what you're doing, you don't know if you're getting where you want to go. You don't even know if you're on the right path. Monitoring is a key part of proactive, precautionary and preventative practice. It helps to reappraise what is achievable, assessing the appropriateness and effectiveness of diagnoses and treatments in efforts to improve condition. Analyses of trajectories of change are required to determine if a threshold condition or tipping point beckons, identifying likely timeframes and consequences, and what can be done about it. In relation to the health analogy, routine check-ups are an important part of medical practice. If something doesn't look or feel right, it's a good idea to act on it, rather than wait.

Failure to learn from the past is a good way to keep making the same mistakes. It pays to respond effectively to lessons learnt, considering the full range of social, cultural, environmental and economic implications. Often, lessons from failures are just as instructive as successes, as they can point to important misconceptions (Bernhardt and Palmer 2011). Admitting failure is a key step in efforts to achieve better outcomes, documenting what does not work and why. In some instances, changes in mindset and practice are a critical measure of performance.

In socio-economic terms, monitoring of river health measures the viability and maintenance of resource and service provision. Such relations must be viewed alongside cultural connections and interactions with the river, framed as measures of wellbeing. In environmental terms, monitoring river health measures the vitality, integrity and functionality of a river, typically appraised relative to interpretations of the best attainable state. Ultimately, the collective summary of these attributes determines the health of the river in relation to things that matter in a given context. In *Finding the Voice of the River*, this reflects measures of the extent to which the river is able to express itself as a living entity, such that its rights are maintained and protected.

Ongoing maintenance is required to respond effectively to changing circumstances. Maintenance works best when it isn't noticed—it simply happens. This echoes the imperative for check-ups and follow-up attention in personal healthcare, especially following a treatment, applying follow-up tests as required. If something doesn't feel right, it probably isn't.

Carefully structured approaches to monitoring and learning balance formal processes of legislative requirements and reporting procedures with locally engaged practices such as citizen science initiatives and community monitoring programmes to appraise the health of the river system as a whole. Representative monitoring sites guide the transferability of insight from one situation to another. Local learning and knowledge-acquisition events can support these endeavours through practical, hands-on activities such as water clarity tests, applications of rapid assessment techniques, bio-blitzes (community-based collection of plants and animals) and social (river) labs. Such endeavours bring participants together, sharing river stories and learning from each other. Structured (carefully scaffolded) approaches to enquiry and quality control compile and communicate catchment-specific information as recurrently updated 'living' information bases. Effective networks support transfer of information, reducing prospects for loss of institutional memory.

Professional development, communication and education roles entail effective mentorship and succession planning programmes to support productive relationships. Apprenticeship schemes based on authentic learning experiences facilitate the acquisition and use of knowledge, concepts and skills in a context-appropriate manner, helping apprentices to become practitioners, rather than merely learning about practice (Dennen and Burner 2008). Many new careers are being created in this process, including roles as coordinators, negotiators, facilitators, intermediaries and communicators. Other roles are not yet imagined, let alone implemented. For example, it is yet to be seen how local authorities and consulting groups will maximize the potential presented for meaningful uptake of place-based understandings generated through applications of emerging technologies in monitoring programmes and measurement techniques. A commitment to ongoing learning is required to support these endeavours.

When conducted effectively, social learning is collective commitment. The iterative process of feeding the feedback loops is an important part of the information age, openly communicating what is happening and why. Meaningful documentation and reporting accompany effective information and knowledge management. Questioning persistently, testing understanding, checking reliability and recurrently updating information bases are critical ingredients in efforts to monitor, evaluate and reappraise.

4.7   CLOSING REFLECTIONS: TRANSITIONING FROM SOLASTALGIA TO SENTIENCE

Wer will dass die welt so bleibt
Wis sie ist der will nicht dass sie bleibt.
He who wants the world to remain as it is
Doesn't want it to remain at all.
Erich Fried (1921–1988)—Art piece on the Berlin Wall

Although Planet Earth has experienced several cataclysmic extinction events, modern times represent the first instance in which an affected party has played a conscious role in the process. In a sense, humanity is 'Playing God' in the Anthropocene, creating particular biological beneficiaries in response to factors such as climate and landscape changes, altered biogeo-chemical process regimes and blending or hybridization of species (Monbiot 2013; Thomas 2017). Such moves perpetuate an evolutionary history of winners and losers, reframed in this instance within a particular image. Increasingly, analytic intelligence and human capacity to observe and predict are being replaced by synthetic intelligence, predicting the future by changing it as human activities redesign, recode and reinvent nature through processes in which engineering recapitulates evolution (Church and Regis 2012). In terms of river futures, key differences are engendered if efforts are framed as engineering exploits to control the world relative to respectful societal relations to the more-than-human world, living with rivers as living entities (Box 4.10).

Box 4.10 Technofix Realities of Future Natures
Disneyland creates an adventurous world that is supposedly removed from the day to day realities of our lives. Many activities focus upon relations to nature, celebrating diversity through fun-laden adventures in 'wild' places. Yet it's a bug-free experience, as pesticides sanitize the air and soil, killing off insects and other forms of life. In such celebrations of nature, we remove ourselves from it. Perhaps it is not so strange that our kids didn't want to hear their dad's reflections on

such matters, as he was supposed to be focussing his attentions on rediscovering his inner spirit of adventure.

Technocentric futures sometimes convey alarming prospects, to me at least. The Sky River Project proposes to create a 'river in the sky' in western China. On first hearing of this proposal at a book launch in Xining in 2016, my instantaneous response was to remind the audience of lessons learnt from working with nature in the Daoist-inspired irrigation scheme at Dujiangyan in nearby Sichuan Province—a scheme that works with the distinctive behavioural traits of an anabranching river and has been continuously operating for more than 2400 years, relative to controlling the lower Yellow River at Sanmenxia in the 1950s, where the dam infilled with sediment in a matter of years following closure.

The Singularity beckons as Artificial Intelligence becomes brainier than us (Kurzweil 2010), wherein laws of intelligent design replace laws of natural selection (Harari 2014). Harari (2016) identifies three primary steps in the transitional process wherein biotechnology is integrated into our bodies, such that human-machine coevolution merges physical and virtual reality. First, bioengineering entails biological changes to our DNA, brain and body, wherein our organic beings extend beyond *Homo sapiens*. The second step moves beyond the organic realm, incorporating bionic (inorganic) parts of cyborgs, such as hearing aids, pacemakers or new joints, and prospectively bionic eyes or nano-robots that attend to our immune system, eventually modifying our brains to work together with computers in a combined thought process. The final step envisages a transition from organic to non-organic entities, transforming notions of human identity as we create new lifeforms—entities with consciousness, feelings and emotions. Profound ethical considerations come to the fore as we scope the emergence of a new species, sometimes referred to as *Homo evolutis*, posthuman, transhuman, parahuman or H+: "Why and how will we teach our robotic or H+ descendants about emotions or morality?" (Church and Regis 2012, p. 253). Inclusive or exclusive access to emerging technologies raises the prospect of biological inequality—a super-human elite. As postulated by Haig (2018, p. 41), will we become immortal and happy cyborgs, or are we destined to become sentient robots? Are we on the brink of a dystopian cyberpunk future, widening our moral circle to include any sentient artificial beings we could create? In such a world, who writes the algorithms to define what constitutes a healthy river?

So, what does this wandering commentary on societal relations to rivers tell us about the prospects for river futures? It highlights the importance of working together, developing a shared commitment to living with rivers that builds upon emotional connections to the more-than-human world. Engaged relationships lie at the heart of a culture and duty of care, acting ethically, with passion. Equitable approaches to social and environmental justice emphasize concerns for all rivers, everywhere, rather than management projects along a few flagship rivers. The tyranny of small thinking diminishes such prospects, fuelling our preoccupation to look after bits (personal property, things we own), hindering collective efforts to look after the Earth as a whole (the commons).

Key lessons to emerge from this book include:

1. Listen to the river and learn to work with it. Echoing the land ethic espoused by Aldo Leopold (1949), humans are an interconnected part of the ecosystem and little can be achieved by fighting nature— as part of it, humanity must work with it (UN 2018).
2. Respect diversity, and the catchment-specific values of rivers. Everywhere is special in its own way, so it pays to identify, protect and enhance place-based values and things that matter.
3. Work collaboratively and collectively.
4. Stop doing stupid things—things that we know do not work. Destruction is easy, but recovery is often exceptionally difficult and frequently impossible such that there is no going back! Once path dependencies are set, they are very difficult to revoke.
5. Act proactively, carefully applying the precautionary principle. Protect things that matter (prevention, not cure), strategically addressing threatening processes (a stitch in time).
6. Work at the landscape scale through a whole of catchment approach from the mountains to the sea, tackling issues at source, before they become a bigger problem elsewhere. Upscale local- and reach-scale interventions to enhance collective impacts, allowing the voice of the river to ring loud and true.
7. Experiment productively, questioning persistently. If something doesn't make sense, don't do it.
8. Learn to live with, and learn from, inherent uncertainties. Adaptivity and flexibility are vital in living with (as a part of) emergent systems, remembering that sometimes it pays to be different (Box 4.11).

> **Box 4.11 Daring to Be Different**
> Sometimes it pays to be true to self, fight adversity and step outside the box. Here are a few examples from my professional career:
>
> - I was advised not to write books as they are no longer valued in academic terms, yet books have made my career.
> - I was told that the title of this book was inappropriate as it does not fit with prescribed marketing programmes and the functionality of Search Engines.
> - In presenting an invited talk at the NZ Pavilion during the World Expo in Shanghai (2008) in a session on creativity and innovation, I looked down from the pedestal upon a sea of black jackets—remarkable commitment to uniformity in practice—thankfully, I'd chosen to wear my tan jacket that day.
> - One of the biggest mistakes in my career, when I didn't trust my instinct, was the decision to trade mark the River Styles® Framework in the quest for quality control. Concerns for copyright, protection of intellectual property and ownership of ideas now clash massively with the creative commons and an open-source world. We live and learn, hopefully.

It's a very slippery slope from solastalgia to sentience, as we negotiate societal relationships to rivers, and to each other. In contemplating prospective futures of the cosmic zoo, are we hopeful, envisaging and enacting the world as it could be, or do we act with fear and lethargy, despairing for the world as it is doomed to be (Schulze-Makuch and Bains 2017)? Is that glass half full, or is it half empty? Are we ready to extend sentient relationships to the more-than-human world, experiencing and sharing metaphysical qualities of all things that require respect and care? 'Finding the Voice of the River' is up to us. Perhaps, with time, we will truly warrant the name that was given to our species, *Homo sapiens*, wherein we act with wisdom. At least we have a choice, for now …

References

Albrecht, G., Sartore, G. M., Connor, L., Higginbotham, N., Freeman, S., Kelly, B., … Pollard, G. (2007). Solastalgia: The distress caused by environmental change. *Australasian Psychiatry, 15*(Suppl 1), S95–S98.

Berkes, F. (1999). *Sacred Ecology. Traditional Ecological Knowledge and Resource Management*. London: Routledge.

Bernhardt, E. S., & Palmer, M. A. (2011). River restoration: The fuzzy logic of repairing reaches to reverse catchment scale degradation. *Ecological Applications, 21*(6), 1926–1931.

Bishop, R., & Glynn, T. (2000). Kaupapa Maori messages for the mainstream. *Set, 1*, 4–7.

Bliss, J. C., & Fischer, A. P. (2011). Toward a political ecology of ecosystem restoration. In D. Egan, E. E. Hjerpe, & J. Abrams (Eds.), *Human Dimensions of Ecological Restoration* (pp. 135–148). Washington, DC: Island Press.

Blue, B. (2018). What's wrong with healthy rivers? Promise and practice in the search for a guiding ideal of freshwater management. *Progress in Physical Geography, 42*, 462–477.

Blue, B., & Brierley, G. (2016). 'But what do you measure?' Prospects for a constructive critical physical geography. *Area, 48*(2), 190–197.

Bodin, Ö., & Crona, B. I. (2009). The role of social networks in natural resource governance: What relational patterns make a difference? *Global Environmental Change, 19*(3), 366–374.

Bosselmann, K. (2011). From reductionist environmental law to sustainability law. In P. Burton (Ed.), *Exploring Wild Law: The Philosophy of Earth Jurisprudence* (pp. 204–213). Adelaide: Wakefield Press.

Brierley, G. J., & Fryirs, K. A. (2016). The use of evolutionary trajectories to guide 'moving targets' in the management of river futures. *River Research and Applications, 32*(5), 823–835.

Brierley, G., Fryirs, K., Cullum, C., Tadaki, M., Huang, H. Q., & Blue, B. (2013). Reading the landscape: Integrating the theory and practice of geomorphology to develop place-based understandings of river systems. *Progress in Physical Geography, 37*(5), 601–621.

Brierley, G., Reid, H., Fryirs, K., & Trahan, N. (2010). What are we monitoring and why? Using geomorphic principles to frame eco-hydrological assessments of river condition. *Science of the Total Environment, 408*(9), 2025–2033.

Budiansky, S. (1995). *Nature's Keepers. The New Science of Nature Management.* London: Weidenfeld & Nicolson.

Capra, F. (1982). *The Turning Point. Science, Society and the Rising Culture.* New York: Simon and Schuster (Bantam Paperback, 1983).

Church, G., & Regis, E. (2012). *Regenesis. How Synthetic Biology Will Reinvent Nature and Ourselves.* New York: Basic Books.

Comby, E., Le Lay, Y. F., & Piégay, H. (2019). Power and changing riverscapes: The socioecological fix and newspaper discourse concerning the Rhône River (France) since 1945. *Annals of the American Association of Geographers.* https://doi.org/10.1080/24694452.2019.1580134

Country, B., Wright, S., Suchet-Pearson, S., Lloyd, K., Burarrwanga, L., Ganambarr, R., ... Sweeney, J. (2016). Co-becoming Bawaka: Towards a relational understanding of place/space. *Progress in Human Geography, 40*(4), 455–475.

Cronon, W. (1992). A place for stories: Nature, history, and narrative. *Journal of American History, 78*(4), 1347–1376.

Dawkins, R. (2017). *Science in the Soul.* London: Transworld.

Dennen, V. P., & Burner, K. J. (2008). The cognitive apprenticeship model in educational practice. In J. M. Spector, M. D. Merrill, J. Merriënboer, & M. P. van Driscoll (Eds.), *Handbook of Research on Educational Communications and Technology* (3rd ed., pp. 425–439). New York: Taylor and Francis.

Dicks, H. (2014). Aldo Leopold and the ecological imaginary: The balance, the pyramid, and the round river. *Environmental Philosophy, 11*(2), 175–209.

Egan, D., Hjerpe, E. E., & Abrams, J. (2011). Why people matter in ecological restoration. In D. Egan, E. E. Hjerpe, & J. Abrams (Eds.), *Human Dimensions of Ecological Restoration* (pp. 1–19). Washington, DC: Island Press.

Elosegi, A., Gessner, M. O., & Young, R. G. (2017). River doctors: Learning from medicine to improve ecosystem management. *Science of the Total Environment, 595,* 294–302.

Flannery, T. F. (2010). *Here on Earth. A Natural History of the Planet.* New York: Atlantic Monthly Press.

Fryirs, K. A. (2015). Developing and using geomorphic condition assessments for river rehabilitation planning, implementation and monitoring. *Wiley Interdisciplinary Reviews: Water, 2*(6), 649–667.

Fryirs, K. A., & Brierley, G. J. (2016). Assessing the geomorphic recovery potential of rivers: Forecasting future trajectories of adjustment for use in management. *Wiley Interdisciplinary Reviews: Water, 3*(5), 727–748.

Fryirs, K. A., Brierley, G. J., Hancock, F., Cohen, T. J., Brooks, A. P., Reinfelds, I., ... Raine, A. (2018). Tracking geomorphic recovery in process-based river management. *Land Degradation & Development, 29*(9), 3221–3244.

Fryirs, K. A., Wheaton, J., Bizzi, S., Williams, R., & Brierley, G. J. (2019). To plug-in or not to plug-in? Geomorphic analysis of rivers using the River Styles Framework in an era of big data acquisition and automation. *Wiley Interdisciplinary Reviews: Water,* e1372.

Gaventa, J. (2006). Finding the spaces for change: A power analysis. *IDS Bulletin, 37*(6), 23–33.

Haig, M. (2018). *Notes on a Nervous Planet.* Edinburgh: Canongate.

Harari, Y. N. (2014). *Sapiens: A Brief History of Humankind.* London: Random House.

Harari, Y. N. (2016). *Homo Deus: A Brief History of Tomorrow.* London: Random House.

Harari, Y. N. (2018). *21 Lessons for the 21st Century.* London: Random House.

Hillman, M. (2005). Justice in river management: Community perceptions from the Hunter Valley, New South Wales. *Geographical Research, 43*(2), 153–161.

Hillman, M. (2006). Situated justice in environmental decision-making: Lessons from river management in Southeastern Australia. *Geoforum, 37*(5), 695–707.

Hillman, M. (2009). Integrating knowledge: The key challenge for a new paradigm in river management. *Geography Compass, 3*(6), 1988–2010.

Huitema, D., Mostert, E., Egas, W., Moellenkamp, S., Pahl-Wostl, C., & Yalcin, R. (2009). Adaptive water governance: Assessing the institutional prescriptions of adaptive (co-) management from a governance perspective and defining a research agenda. *Ecology and Society, 14*(1).

Johnson, J. T., Howitt, R., Cajete, G., Berkes, F., Louis, R. P., & Kliskey, A. (2016). Weaving Indigenous and sustainability sciences to diversify our methods. *Sustainability Science, 11*(1), 1–11.

Kurzweil, R. (2010). *The Singularity Is Near*. London: Gerald Duckworth & Co.

Larsen, S. C., & Johnson, J. T. (2017). *Being Together in Place: Indigenous Co-Existence in a More Than Human World*. Minneapolis, MN: University of Minnesota Press.

Leopold, A. (1949). *A Sand County Almanac and Sketches Here and There*. Oxford: Oxford University Press.

Longley, A., Fitzpatrick, K., Martin, R., Brown, C., Šunde, C., Ehlers, C., ... Waghorn, K. (2013). Imagining a fluid city. *Qualitative Inquiry, 19*(9), 736–740.

Mahood, K. (2016). *Position Doubtful. Mapping Landscapes and Memories*. Melbourne, VIC: Scribe.

Mika, S., Hoyle, J., Kyle, G., Howell, T., Wolfenden, B., Ryder, D., ... Fryirs, K. (2010). Inside the "black box" of river restoration: Using catchment history to identify disturbance and response mechanisms to set targets for process-based restoration. *Ecology and Society, 15*(4).

Mitchell, D. (2003). Cultural landscapes: Just landscapes or landscapes of justice? *Progress in Human Geography, 27*(6), 787–796.

Monbiot, G. (2013). *Feral: Searching for Enchantment on the Frontiers of Rewilding*. London: Penguin.

Morin, K. M. (2009). Landscape: Representing and interpreting the world. In N. J. Clifford, S. L. Holloway, S. P. Rice, & G. Valentine (Eds.), *Key Concepts in Geography* (2nd ed., pp. 286–299). Los Angeles: Sage.

Mould, S. A., Fryirs, K., & Howitt, R. (2018). Practicing sociogeomorphology: Relationships and dialog in river research and management. *Society & Natural Resources, 31*(1), 106–120.

Nassauer, J. I. (1995). Culture and changing landscape structure. *Landscape Ecology, 10*(4), 229–237.

Norgaard, R. B. (2006). *Development Betrayed: The End of Progress and a Co-Evolutionary Revisioning of the Future*. London: Routledge.

Nucitelli, D. (2017, December 27). Fake news is a threat to humanity, but scientists may have a solution. *The Guardian*.

O'Neil, C. (2016). *Weapons of Math Destruction: How Big Data Increases Inequality and Threatens Democracy*. London: Penguin.

O'Neill, J., Holland, A., & Light, A. (2008). *Environmental Values*. London: Routledge.

Oreskes, N., & Conway, E. M. (2011). *Merchants of Doubt: How a Handful of Scientists Obscured the Truth on Issues from Tobacco Smoke to Global Warming*. London: Bloomsbury Publishing.

Palmer, T. (1994). *Lifelines. The Case for River Conservation*. Washington, DC: Island Press.

Palmer, M. A., Bernhardt, E. S., Allan, J. D., Lake, P. S., Alexander, G., Brooks, S., … Galat, D. L. (2005). Standards for ecologically successful river restoration. *Journal of Applied Ecology, 42*(2), 208–217.

Plumwood, V. (2002). *Feminism and the Mastery of Nature*. London: Routledge.

Relph, E. (1976). *Place and Placelessness*. London: Pion.

Rogers, K. H. (2006). The real river management challenge: Integrating scientists, stakeholders and service agencies. *River Research and Applications, 22*(2), 269–280.

Salmond, A. (2017). *Tears of Rangi: Experiments Across Worlds*. Auckland: Auckland University Press.

Schama, S. (1995). *Landscape and Memory*. New York: Knopf.

Schmidt, J. C., Webb, R. H., Valdez, R. A., Marzolf, G. R., & Stevens, L. E. (1998). Science and values in river restoration in the Grand Canyon: There is no restoration or rehabilitation strategy that will improve the status of every riverine resource. *BioScience, 48*(9), 735–747.

Schulze-Makuch, D., & Bains, W. (2017). *The Cosmic Zoo: Complex Life on Many Worlds*. New York: Springer.

Spink, A., Fryirs, K., & Brierley, G. (2009). The relationship between geomorphic river adjustment and management actions over the last 50 years in the upper Hunter catchment, NSW, Australia. *River Research and Applications, 25*(7), 904–928.

Tadaki, M., & Sinner, J. (2014). Measure, model, optimise: Understanding reductionist concepts of value in freshwater governance. *Geoforum, 51*, 140–151.

Thomas, C. D. (2017). *Inheritors of the Earth*. London: Allen Lane, Penguin Random House.

Tsing, A. L. (2004). *Friction*. Princeton, NJ: Princeton University Press.

Tuan, Y. F. (1974). *Topophilia: A Study of Environmental Perceptions, Attitudes, and values*. New York: Columbia University Press.

Tuan, Y. F. (1977). *Space and Place: The Perspective of Experience*. Minneapolis, MN: University of Minnesota Press.

UN (WWAP (United Nations World Water Assessment Programme)/UN-Water). (2018). *The United Nations World Water Development Report 2018: Nature-Based Solutions for Water.* Paris: UNESCO.

Vugteveen, P., Leuven, R. S. E. W., Huijbregts, M. A. J., & Lenders, H. J. R. (2006). Redefinition and elaboration of river ecosystem health: Perspective for river management. *Hydrobiologia, 565,* 289–308.

Walker, W. E., Loucks, D. P., & Carr, G. (2015). Social responses to water management decisions. *Environmental Processes, 2*(3), 485–509.

White, R. (1990). Environmental history, ecology, and meaning. *The Journal of American History, 76*(4), 1111–1116.

Wilcock, D., Brierley, G., & Howitt, R. (2013). Ethnogeomorphology. *Progress in Physical Geography, 37*(5), 573–600.

Wilkinson, J. L., Hooda, P. S., Swinden, J., Barker, J., & Barton, S. (2018). Spatial (bio) accumulation of pharmaceuticals, illicit drugs, plasticisers, perfluorinated compounds and metabolites in river sediment, aquatic plants and benthic organisms. *Environmental Pollution, 234,* 864–875.

Wylie, J. (2007). *Landscape.* Abingdon: Routledge.

Yu, G. A., Huang, H. Q., Wang, Z., Brierley, G., & Zhang, K. (2012). Rehabilitation of a debris-flow prone mountain stream in southwestern China—Strategies, effects and implications. *Journal of Hydrology, 414,* 231–243.

References

Ackroyd, P. (2007). *Thames: Sacred River*. New York: Doubleday.

Acreman, M. (2016). Environmental flows—Basics for novices. *Wiley Interdisciplinary Reviews: Water, 3*(5), 622–628.

Albrecht, G., Sartore, G. M., Connor, L., Higginbotham, N., Freeman, S., Kelly, B., ... Pollard, G. (2007). Solastalgia: The distress caused by environmental change. *Australasian Psychiatry, 15*(Suppl 1), S95–S98.

Armitage, D., de Loe, R., & Plummer, R. (2012). Environmental governance and its implications for conservation practice. *Conservation Letters, 5*(4), 245–255.

Aronson, J., & Alexander, S. (2013). Ecosystem restoration is now a global priority: Time to roll up our sleeves. *Restoration Ecology, 21*(3), 293–296.

Arthington, A. H. (2012). *Environmental Flows: Saving Rivers in the Third Millennium* (Vol. 4). Berkeley, CA: University of California Press.

Barclay, L., Gifford, T., & Linke, S. (2018). River listening: Acoustic ecology and aquatic bioacoustics. *Leonardo, 51*(3), 298–299.

Beattie, J., & Morgan, R. (2017). Engineering Edens on this 'rivered earth'? A review article on water management and hydro-resilience in the British Empire, 1860–1940s. *Environment and History, 23*(1), 39–63.

Benda, L. E., Poff, N. L., Tague, C., Palmer, M. A., Pizzuto, J., Cooper, S., ... Moglen, G. (2002). How to avoid train wrecks when using science in environmental problem solving. *Bioscience, 52*(12), 1127–1136.

Berkes, F. (1999). *Sacred Ecology. Traditional Ecological Knowledge and Resource Management*. London: Routledge.

© The Author(s) 2020
G. J. Brierley, *Finding the Voice of the River*,
https://doi.org/10.1007/978-3-030-27068-1

Bernhardt, E. S., & Palmer, M. A. (2011). River restoration: The fuzzy logic of repairing reaches to reverse catchment scale degradation. *Ecological Applications, 21*(6), 1926–1931.

Bernhardt, E. S., Palmer, M. A., Allan, J. D., Alexander, G., Barnas, K., Brooks, S., ... Galat, D. (2005). Synthesizing US river restoration efforts. *Science, 308*(5722), 636–637.

Best, J. (2019). Anthropogenic stresses on the world's big rivers. *Nature Geoscience, 12*, 7–21.

Biron, P. M., Buffin-Bélanger, T., Larocque, M., Choné, G., Cloutier, C. A., Ouellet, M. A., ... Eyquem, J. (2014). Freedom space for rivers: A sustainable management approach to enhance river resilience. *Environmental Management, 54*(5), 1056–1073.

Bishop, R., & Glynn, T. (2000). Kaupapa Maori messages for the mainstream. *Set, 1*, 4–7.

Blackstock, M. (2001). Water: A first nations' spiritual and ecological perspective. *BC Journal of Ecosystems and Management, 1*, 2–14.

Bliss, J. C., & Fischer, A. P. (2011). Toward a political ecology of ecosystem restoration. In D. Egan, E. E. Hjerpe, & J. Abrams (Eds.), *Human Dimensions of Ecological Restoration* (pp. 135–148). Washington, DC: Island Press.

Blue, B. (2018). What's wrong with healthy rivers? Promise and practice in the search for a guiding ideal of freshwater management. *Progress in Physical Geography, 42*, 462–477.

Blue, B., & Brierley, G. (2016). 'But what do you measure?' Prospects for a constructive critical physical geography. *Area, 48*(2), 190–197.

Bodin, Ö., & Crona, B. I. (2009). The role of social networks in natural resource governance: What relational patterns make a difference? *Global Environmental Change, 19*(3), 366–374.

Bosselmann, K. (2008). *The Principle of Sustainability: Transforming Law and Governance.* London: Routledge.

Bosselmann, K. (2011). From reductionist environmental law to sustainability law. In P. Burton (Ed.), *Exploring Wild Law: The Philosophy of Earth Jurisprudence* (pp. 204–213). Adelaide: Wakefield Press.

Bouleau, G. (2014). The co-production of science and waterscapes: The case of the Seine and the Rhône Rivers, France. *Geoforum, 57*, 248–257.

Boulton, A. J., Piégay, H., & Sanders, M. D. (2008). Turbulence and train wrecks: Using knowledge strategies to enhance the application of integrative river science in effective river management. In G. J. Brierley & K. A. Fryirs (Eds.), *River Futures: An Integrative Scientific Approach to River Repair* (pp. 28–39). Washington, DC: Island Press.

Bowler, P. J. (1992). *The Fontana History of the Environmental Sciences.* London: Fontana.

Boyd, D. R. (2017). *The Rights of Nature.* Toronto, ON: ECW Press.

Brierley, G. J., & Fryirs, K. A. (2005). *Geomorphology and River Management: Applications of the River Styles Framework.* Chichester: John Wiley & Sons.

Brierley, G. J., & Fryirs, K. A. (Eds.). (2008). *River Futures.* Washington, DC: Island Press.

Brierley, G., & Fryirs, K. (2009). Don't fight the site: Three geomorphic considerations in catchment-scale river rehabilitation planning. *Environmental Management, 43*(6), 1201–1218.

Brierley, G. J., & Fryirs, K. A. (2016). The use of evolutionary trajectories to guide 'moving targets' in the management of river futures. *River Research and Applications, 32*(5), 823–835.

Brierley, G., Fryirs, K., Cullum, C., Tadaki, M., Huang, H. Q., & Blue, B. (2013). Reading the landscape: Integrating the theory and practice of geomorphology to develop place-based understandings of river systems. *Progress in Physical Geography, 37*(5), 601–621.

Brierley, G., Reid, H., Fryirs, K., & Trahan, N. (2010). What are we monitoring and why? Using geomorphic principles to frame eco-hydrological assessments of river condition. *Science of the Total Environment, 408*(9), 2025–2033.

Brierley, G., Tadaki, M., Hikuroa, D., Blue, B., Šunde, C., Tunnicliffe, J., & Salmond, A. (2018). A geomorphic perspective on the rights of the river in Aotearoa New Zealand. *River Research and Applications.* https://doi.org/10.1002/rra.3343

Brower, D., & Chapple, S. (1995). *Let the Mountains Speak, Let the Rivers Run: A Call to Those Who Would Save the Earth.* New York: Harper Collins.

Budiansky, S. (1995). *Nature's Keepers. The New Science of Nature Management.* London: Weidenfeld & Nicolson.

Buffin-Bélanger, T., Biron, P. M., Larocque, M., Demers, S., Olsen, T., Choné, G., … Eyquem, J. (2015). Freedom space for rivers: An economically viable river management concept in a changing climate. *Geomorphology, 251,* 137–148.

Capra, F. (1982). *The Turning Point. Science, Society and the Rising Culture.* New York: Simon and Schuster (Bantam Paperback, 1983).

Carrizo, S. F., Jähnig, S. C., Bremerich, V., Freyhof, J., Harrison, I., He, F., … Darwall, W. (2017). Freshwater megafauna: Flagships for freshwater biodiversity under threat. *BioScience, 67*(10), 919–927.

Carson, R. (1962). *Silent Spring.* Boston: Houghton Mifflin Harcourt.

Chakraborty, A., & Chakraborty, S. (2018). Rivers as socioecological landscapes. Chap. 2 in Cooper, M., Chakraborty, A., & Chakraborty, S. (Eds.), *Rivers and Society: Landscapes, Governance and Livelihoods.* London: Routledge.

Chamovitz, D. (2012). *What a Plant Knows: A Field Guide to the Senses.* New York: Scientific American/Farrar, Straus and Giroux.

Chapin, F. S., Carpenter, S. R., Kofinas, G. P., Folke, C., Abel, N., Clark, W. C., … Berkes, F. (2010). Ecosystem stewardship: Sustainability strategies for a rapidly changing planet. *Trends in Ecology & Evolution, 25*(4), 241–249.

Chapron, G., Epstein, Y., & López-Bao, J. V. (2019). A rights revolution for nature. *Science, 363*(6434), 1392–1393.

Charpleix, L. (2018). The Whanganui River as Te Awa Tupua: Place-based law in a legally pluralistic society. *The Geographical Journal, 184*(1), 19–30.

Church, G., & Regis, E. (2012). *Regenesis. How Synthetic Biology Will Reinvent Nature and Ourselves*. New York: Basic Books.

Cioc, M. (2002). *The Rhine: An Eco-Biography*. Seattle, WA: University of Washington Press.

Cohen, A. (2012). Rescaling environmental governance: Watersheds as boundary objects at the intersection of science, neoliberalism, and participation. *Environment and Planning A, 44*(9), 2207–2224.

Collard, R. C., Dempsey, J., & Sundberg, J. (2015). A manifesto for abundant futures. *Annals of the Association of American Geographers, 105*(2), 322–330.

Collier, K. J. (2017). Measuring river restoration success: Are we missing the boat? *Aquatic Conservation: Marine and Freshwater Ecosystems, 27*(3), 572–577.

Comby, E., Le Lay, Y. F., & Piégay, H. (2019). Power and changing riverscapes: The socioecological fix and newspaper discourse concerning the Rhône River (France) since 1945. *Annals of the American Association of Geographers*. https://doi.org/10.1080/24694452.2019.1580134

Cook, C., & Bakker, K. (2012). Water security: Debating an emerging paradigm. *Global Environmental Change, 22*(1), 94–102.

Country, B., Suchet-Pearson, S., Wright, S., Lloyd, K., Tofa, M., Burarrwanga, L., … Maymuru, D. (2019). Bunbum ga dhä-yu t agum: To make it right again, to remake. *Social & Cultural Geography*. https://doi.org/10.1080/1464936 5.2019.1584825

Country, B., Wright, S., Suchet-Pearson, S., Lloyd, K., Burarrwanga, L., Ganambarr, R., … Sweeney, J. (2016). Co-becoming Bawaka: Towards a relational understanding of place/space. *Progress in Human Geography, 40*(4), 455–475.

Cronon, W. (1992). A place for stories: Nature, history, and narrative. *Journal of American History, 78*(4), 1347–1376.

Cronon, W. (1996). The trouble with wilderness: Or, getting back to the wrong nature. *Environmental History, 1*(1), 7–28.

Cruikshank, J. (2014). *Do Glaciers Listen? Local Knowledge, Colonial Encounters, and Social Imagination*. Vancouver, BC: UBC Press.

Dart, R. A. (1940). Recent discoveries bearing on human history in southern Africa. *The Journal of the Royal Anthropological Institute of Great Britain and Ireland, 70*(1), 13–27.

Dartnell, L. (2018). *Origins. How the Earth Made Us*. London: The Bodley Head.

Davison, N. (2019, May 30). The Anthropocene epoch: Have we entered a new phase of planetary history. *The Guardian*.

Dawkins, R. (2017). *Science in the Soul*. London: Transworld.

Dempsey, J. (2016). *Enterprising Nature: Economics, Markets, and Finance in Global Biodiversity Politics*. Chichester: John Wiley & Sons.

Dennen, V. P., & Burner, K. J. (2008). The cognitive apprenticeship model in educational practice. In J. M. Spector, M. D. Merrill, J. Merriënboer, & M. P. van Driscoll (Eds.), *Handbook of Research on Educational Communications and Technology* (3rd ed., pp. 425–439). New York: Taylor and Francis.

Diamond, J. M. (1997). *Guns, Germs and Steel: A Short History of Everybody for the Last 13,000 Years*. New York: W.H. Norton.

Diamond, J. (2005). *Collapse: How Societies Choose to Fail or Succeed*. New York: Penguin.

Diamond, J. (2013). *The World Until Yesterday: What Can We Learn from Traditional Societies?* New York: Penguin.

Dicks, H. (2014). Aldo Leopold and the ecological imaginary: The balance, the pyramid, and the round river. *Environmental Philosophy, 11*(2), 175–209.

Downs, P., & Gregory, K. (2004). *River Channel Management: Towards Sustainable Catchment Hydrosystems*. London: Routledge.

Doyle, M. (2018). *The Source: How Rivers Made America and America Remade Its Rivers*. New York: Norton.

Drew, G. (2013). Why wouldn't we cry? Love and loss along a river in decline. *Emotion, Space and Society, 6*, 25–32.

Dudley, M. (2017). Muddying the waters: Recreational conflict and rights of use of British rivers. *Water History, 9*(3), 259–277.

Dufour, S., & Piégay, H. (2009). From the myth of a lost paradise to targeted river restoration: Forget natural references and focus on human benefits. *River Research and Applications, 25*(5), 568–581.

Duncan, D. J. (2001). *My Story as Told by Water*. San Francisco: Sierra Book Clubs.

East, A. E., Pess, G. R., Bountry, J. A., Magirl, C. S., Ritchie, A. C., Logan, J. B., … Liermann, M. C. (2015). Large-scale dam removal on the Elwha River, Washington, USA: River channel and floodplain geomorphic change. *Geomorphology, 228*, 765–786.

Eden, S. (2017). Environmental restoration. In D. Richardson, N. Castree, M. F. Goodchild, A. Kobayashi, W. Liu, & R. A. Marston (Eds.), *The International Encyclopedia of Geography*. https://doi.org/10.1002/9781118786352.wbieg0622

Eden, S., & Tunstall, S. (2006). Ecological versus social restoration? How urban river restoration challenges but also fails to challenge the science–policy nexus in the United Kingdom. *Environment and Planning C: Government and Policy, 24*(5), 661–680.

Egan, D., Hjerpe, E. E., & Abrams, J. (Eds.). (2011a). *Human Dimensions of Ecological Restoration*. Washington, DC: Island Press.

Egan, D., Hjerpe, E. E., & Abrams, J. (2011b). Why people matter in ecological restoration. In D. Egan, E. E. Hjerpe, & J. Abrams (Eds.), *Human Dimensions of Ecological Restoration* (pp. 1–19). Washington, DC: Island Press.

Elliot, R. (1997). *Faking Nature: The Ethics of Environmental Restoration.* London: Routledge.

Ellis, J. (Ed.). (2018). *Water Rites: Reimagining Water in the West.* Calgary: Calgary University Press.

Elosegi, A., Gessner, M. O., & Young, R. G. (2017). River doctors: Learning from medicine to improve ecosystem management. *Science of the Total Environment, 595,* 294–302.

Emery, S. B., Perks, M. T., & Bracken, L. J. (2013). Negotiating river restoration: The role of divergent reframing in environmental decision-making. *Geoforum, 47,* 167–177.

Engels, F. (1910). *Socialism: Utopian and Scientific.* Chicago: Kerr and Co.

Everard, M., & Moggridge, H. L. (2012). Rediscovering the value of urban rivers. *Urban Ecosystems, 15*(2), 293–314.

Everard, M., & Powell, A. (2002). Rivers as living systems. *Aquatic Conservation: Marine and Freshwater Ecosystems, 12*(4), 329–337.

Fairbanks, D. J. (2012). *Evolving: The Human Effect and Why It Matters.* Amherst, NY: Prometheus Books.

Fausch, K. D. (2015). *For the Love of Rivers. A Scientist's Journey.* Corvallis: Oregon State University Press.

Fausch, K. D., Torgersen, C. E., Baxter, C. V., & Li, H. W. (2002). Landscapes to riverscapes: Bridging the gap between research and conservation of stream fishes. *BioScience, 52*(6), 1–16.

Fisher, S. G. (1997). Creativity, idea generation, and the functional morphology of streams. *Journal of the North American Benthological Society, 16*(2), 305–318.

Flannery, T. F. (2010). *Here on Earth. A Natural History of the Planet.* New York: Atlantic Monthly Press.

Florsheim, J. L., Mount, J. F., & Chin, A. (2008). Bank erosion as a desirable attribute of rivers. *BioScience, 58*(6), 519–529.

Fox, C. A., Reo, N. J., Turner, D. A., Cook, J., Dituri, F., Fessell, B., ... Turner, A. (2017). "The river is us; the river is in our veins": Re-defining river restoration in three Indigenous communities. *Sustainability Science, 12*(4), 521–533.

Freudenburg, W. R., Frickel, S., & Gramling, R. (1995). Beyond the nature/society divide: Learning to think about a mountain. *Sociological Forum, 10*(3), 361–392.

Friedman, T. L. (2006). *The World Is Flat: The Globalized World in the Twenty-First Century.* London: Penguin.

Friends of the Earth. (1972). *Only One Earth. The Care and Maintenance of a Small Planet.* Harmondsworth: Penguin.

Fromm, E. (1964). *The Heart of Man.* New York: Harper & Row.

Fryirs, K. A. (2015). Developing and using geomorphic condition assessments for river rehabilitation planning, implementation and monitoring. *Wiley Interdisciplinary Reviews: Water, 2*(6), 649–667.

Fryirs, K. A., & Brierley, G. J. (2016). Assessing the geomorphic recovery potential of rivers: Forecasting future trajectories of adjustment for use in management. *Wiley Interdisciplinary Reviews: Water, 3*(5), 727–748.

Fryirs, K. A., Brierley, G. J., Hancock, F., Cohen, T. J., Brooks, A. P., Reinfelds, I., ... Raine, A. (2018). Tracking geomorphic recovery in process-based river management. *Land Degradation & Development, 29*(9), 3221–3244.

Fryirs, K. A., Wheaton, J., Bizzi, S., Williams, R., & Brierley, G. J. (2019). To plug-in or not to plug-in? Geomorphic analysis of rivers using the River Styles Framework in an era of big data acquisition and automation. *Wiley Interdisciplinary Reviews: Water*, e1372.

Gaventa, J. (2006). Finding the spaces for change: A power analysis. *IDS Bulletin, 37*(6), 23–33.

Geay, T., Belleudy, P., Gervaise, C., Habersack, H., Aigner, J., Kreisler, A., ... Laronne, J. B. (2017). Passive acoustic monitoring of bed load discharge in a large gravel bed river. *Journal of Geophysical Research: Earth Surface, 122*(2), 528–545.

Ghosh, A. (2016). *The Great Derangement: Climate Change and the Unthinkable.* Chicago: University of Chicago Press.

Gibbs, L. (2014). Freshwater geographies? Place, matter, practice, hope. *New Zealand Geographer, 70*(1), 56–60.

Gilmer, R. (2013). Snail darters and sacred places: Creative application of the Endangered Species Act. *Environmental Management, 52*(5), 1046–1056.

Ginn, F., & Demeritt, D. (2009). Nature: A contested concept. In N. J. Clifford, S. L. Holloway, S. P. Rice, & G. Valentine (Eds.), *Key Concepts in Geography* (2nd ed., pp. 300–311). Los Angeles: Sage.

Gleick, P. H., Heberger, M., & Donnelly, K. (2014). Zombie water projects. In P. H. Gleick (Ed.), *The World's Water* (pp. 123–146). Washington, DC: Island Press.

Gobster, P. H., & Hull, R. B. (Eds.). (2000). *Restoring Nature: Perspectives from the Social Sciences and Humanities.* Washington, DC: Island Press.

Gregory, R. (2014). *Restoring the Life of the Land: Taro Patches in Hawai'i.* Retrieved from http://ecotippingpoints.org/our-stories/indepth/usa-hawaii-taro-agriculture.html

Gregory, C., Brierley, G. J., & Le Heron, R. (2011). Governance spaces for sustainable river management. *Geography Compass, 5*(4), 182–199.

Groenfeldt, D. (2016). Cultural water wars: Power and hegemony in the semiotics of water. In C. M. Ashcraft & T. Mayer (Eds.), *The Politics of Fresh Water: Access, Conflict and Identity* (pp. 143–156). London: Routledge.

Habersack, H., & Piégay, H. (2007). River restoration in the Alps and their surroundings: Past experience and future challenges. *Developments in Earth Surface Processes, 11*, 703–735.

Haeckel, E. H. P.-A. (1866). *Generelle Morphologie der Organismen: Allgemeine Grundzüge der organischen Formen-Wissenschaft, mechanisch begründet durch die von Charles Darwin reformirte Descendenz-Theorie.* Berlin: Druck und Verlag.

Haig, M. (2018). *Notes on a Nervous Planet.* Edinburgh: Canongate.

Hammer, C. (2011). *The River: A Journey Through the Murray-Darling Basin.* Melbourne: Melbourne University Publishing.

Hannam, P. (2019, January 20). 'Cultural water': Indigenous water claims finally on Darling agenda. *Sydney Morning Herald.*

Harari, Y. N. (2014). *Sapiens: A Brief History of Humankind.* London: Random House.

Harari, Y. N. (2016). *Homo Deus: A Brief History of Tomorrow.* London: Random House.

Harari, Y. N. (2018). *21 Lessons for the 21st Century.* London: Random House.

Haraway, D. J. (1981). *Simians, Cyborgs, and Women: The Reinvention of Nature.* London: Free Association Books.

Hardin, G. (1968). The tragedy of the commons. *Science, 162,* 1243–1248.

Harmsworth, G. R., & Awatere, S. (2013). Indigenous Māori knowledge and perspectives of ecosystems. In J. R. Dymond (Ed.), *Ecosystem Services in New Zealand: Conditions and Trends. Manaaki Whenua Press, Landcare Research, Palmerston North* (pp. 274–286). Lincoln: Manaaki Whenua Press.

Harmsworth, G., Awatere, S., & Robb, M. (2016). Indigenous Māori values and perspectives to inform freshwater management in Aotearoa-New Zealand. *Ecology and Society, 21*(4).

Harris, G. P., & Heathwaite, A. L. (2012). Why is achieving good ecological outcomes in rivers so difficult? *Freshwater Biology, 57,* 91–107.

Hartwig, L. D., Jackson, S., & Osborne, N. (2018). Recognition of Barkandji water rights in Australian settler-colonial water regimes. *Resources, 7*(1), 16.

Harvey, D. (1996). *Justice, Nature and the Geography of Difference.* Malden, MA: Blackwell.

Harvey, D. (2007). *A Brief History of Neoliberalism.* Oxford: Oxford University Press.

He, F., Zarfl, C., Bremerich, V., David, J. N., Hogan, Z., Kalinkat, G., ... Jähnig, S. C. (2019). The global decline of freshwater megafauna. *Global Change Biology.* https://doi.org/10.1111/gcb.14753

He, F., Zarfl, C., Bremerich, V., Henshaw, A., Darwall, W., Tockner, K., & Jähnig, S. C. (2017). Disappearing giants: A review of threats to freshwater megafauna. *Wiley Interdisciplinary Reviews: Water, 4*(3), e1208.

Herman, D. (2016, June 1). Finding lessons on culture and conservation at the end of the road in Kauai. *Smithsonian Magazine.* Retrieved from Smithsonian. com

Hesse, H. (1951). *Siddhartha.* New York: Bantam.

Higgs, E. (2003). *Nature by Design.* Cambridge, MA: MIT Press.

Higgs, E. (2011). Foreword. In D. Egan, E. E. Hjerpe, & J. Abrams (Eds.), *Human Dimensions of Ecological Restoration* (pp. xvii–xxix). Washington, DC: Island Press.

Hikuroa, D., Brierley, G. J., Tadaki, M., Blue, B., & Salmond, A. (2019). Restoring socio-cultural relationships with rivers: Experiments in fluvial pluralism from Aotearoa New Zealand. In Cottet, M., Morandi, B, & Piégay, H. (Eds). Socio-Cultural Perspectives in River Restoration. Wiley (in press).

Hilderbrand, R. H., Watts, A. C., & Randle, A. M. (2005). The myths of restoration ecology. *Ecology and Society, 10*(1).

Hillman, M. (2005). Justice in river management: Community perceptions from the Hunter Valley, New South Wales. *Geographical Research, 43*(2), 153–161.

Hillman, M. (2006). Situated justice in environmental decision-making: Lessons from river management in Southeastern Australia. *Geoforum, 37*(5), 695–707.

Hillman, M. (2009). Integrating knowledge: The key challenge for a new paradigm in river management. *Geography Compass, 3*(6), 1988–2010.

Hillman, M., & Brierley, G. (2005). A critical review of catchment-scale stream rehabilitation programmes. *Progress in Physical Geography, 29*(1), 50–76.

Hillman, M., & Brierley, G. J. (2008). Restoring uncertainty: translating science into management practice. In G. J. Brierley & K. A. Fryirs (Eds.), *River Futures: An Integrative Scientific Approach to River Repair* (pp. 257–272). Washington, DC: Island Press.

Hoefle, S. W. (2016). Além da sociedade-natureza com a mais-que-geografia humana: por uma teoria transdisciplinar de ética ambiental e visão do mundo. In E. S. Sposito, C. A. Silva, J. Sant'anna Neto, & E. S. Melazzo (Eds.), *A diversidade da Geografia brasiliera* (pp. 467–505). Rio de Janeiro: Consequência, UFRJ.

Howitt, R. (2018). Indigenous rights vital to survival. *Nature Sustainability, 1*(7), 339.

Howitt, R., & Suchet-Pearson, S. (2006). Rethinking the building blocks: Ontological pluralism and the idea of 'management'. *Geografiska Annaler: Series B, Human Geography, 88*(3), 323–335.

Huitema, D., Mostert, E., Egas, W., Moellenkamp, S., Pahl-Wostl, C., & Yalcin, R. (2009). Adaptive water governance: Assessing the institutional prescriptions of adaptive (co-) management from a governance perspective and defining a research agenda. *Ecology and Society, 14*(1).

Hussey, K., & Pittock, J. (2012). The energy–water nexus: Managing the links between energy and water for a sustainable future. *Ecology and Society, 17*(1).

Hüttl, R. F., Bens, O., Bismuth, C., Hoechstetter, S., Frede, H.-G., & Kümpel, H.-J. (2016). Introduction: A critical appraisal of major water engineering projects and the need for interdisciplinary approaches. In R. F. Hüttl, O. Bens, C. Bismuth, & S. Hoechstetter (Eds.), *Society-Water-Technology* (pp. 3–9). Cham: Springer.

Jackson, S. (2018). Water and Indigenous rights: Mechanisms and pathways of recognition, representation, and redistribution. *Wiley Interdisciplinary Reviews: Water, 5*(6), e1314.

Jackson, S., & Palmer, L. R. (2015). Reconceptualizing ecosystem services: Possibilities for cultivating and valuing the ethics and practices of care. *Progress in Human Geography, 39*(2), 122–145.

Jähnig, S. C., Lorenz, A. W., Hering, D., Antons, C., Sundermann, A., Jedicke, E., & Haase, P. (2011). River restoration success: A question of perception. *Ecological Applications, 21*(6), 2007–2015.

Johnson, J. T., Howitt, R., Cajete, G., Berkes, F., Louis, R. P., & Kliskey, A. (2016). Weaving Indigenous and sustainability sciences to diversify our methods. *Sustainability Science, 11*(1), 1–11.

Jones, J., Börger, L., Tummers, J., Jones, P., Lucas, M., Kerr, J., … Vowles, A. (2019). A comprehensive assessment of stream fragmentation in Great Britain. *Science of the Total Environment, 673*, 756–762.

Jowett, I. G., & Biggs, B. J. (2009). Application of the 'natural flow paradigm' in a New Zealand context. *River Research and Applications, 25*(9), 1126–1135.

Kareiva, P., Watts, S., McDonald, R., & Boucher, T. (2007). Domesticated nature: Shaping landscapes and ecosystems for human welfare. *Science, 316*(5833), 1866–1869.

Karr, J. R. (1999). Defining and measuring river health. *Freshwater Biology, 41*(2), 221–234.

Katz, E. (1996). The problem of ecological restoration. *Environmental Ethics, 18*(2), 222–224.

Kersten, J. (2013). The enjoyment of complexity: A new political anthropology for the Anthropocene? *RCC Perspectives, 3*, 39–56.

Kim, H., & Jung, Y. (2018). Is Cheonggyecheon sustainable? A systematic literature review of a stream restoration in Seoul. *South Korea. Sustainable Cities and Society, 45*, 59–69.

Kimmerer, R. (2011). Restoration and reciprocity: The contributions of traditional ecological knowledge. In D. Egan, E. E. Hjerpe, & J. Abrams (Eds.), *Human Dimensions of Ecological Restoration* (pp. 257–276). Washington, DC: Island Press.

Kinkaid, E. (2019). "Rights of nature" in translation: Assemblage geographies, boundary objects, and translocal social movements. *Transactions of the Institute of British Geographers.* https://doi.org/10.1111/tran.12303

Knight, C. (2016). *New Zealand's Rivers. An Environmental History* (p. 323). Christchurch: Canterbury University Press.

Kondolf, G. M. (2006). River restoration and meanders. *Ecology and Society, 11*(2).

Kondolf, G. M. (2011). Setting goals in river restoration: When and where can the river "heal itself"? In A. Simon, S. J. Bennett, & J. M. Castro (Eds.), *Stream Restoration in Dynamics Fluvial Systems: Scientific Approaches, Analyses and*

Tools (Geophysical Monograph Series) (Vol. 194, pp. 29–43). Washington, DC: American Geophysical Union.

Kondolf, G. M. (2012). The Espace de Liberté and restoration of fluvial process: When can the river restore itself and when must we intervene? In P. J. Boon & P. J. Raven (Eds.), *River Conservation and Management* (pp. 225–242). Chichester: Wiley.

Kondolf, G. M., & Pinto, P. J. (2017). The social connectivity of urban rivers. *Geomorphology, 277*, 182–196.

Kondolf, G. M., & Yang, C. N. (2008). Planning river restoration projects: Social and cultural dimensions. In S. Darby & D. Sear (Eds.), *River Restoration: Managing the Uncertainty in Restoring Physical Habitat* (pp. 43–60). Chichester: Wiley.

Kothari, A., & Bajpai, S. (2017). We are the river, the river is us. *Economic and Political Weekly, 52*, 103–109.

Kovel, J. S. (2002). *The Enemy of Nature: The End of Capitalism or the End of the World?* London: Zed Books.

Kozacek, C. (2014, October 28). Hawaii River restorations reflect national desire to protect water for public benefit. *Circle of Blue.*

Kurzweil, R. (2010). *The Singularity Is Near.* London: Gerald Duckworth & Co.

Kyle, M. (2013). The 'Four Great Waters' Case: An important expansion of Waiahole Ditch and the Public Trust Doctrine. *U. Denver Water Law Review, 17*, 1–27.

LaBoucane-Benson, P., Gibson, G., Benson, A., & Miller, G. (2012). Are we seeking Pimatisiwin or creating Pomewin? Implications for water policy. *International Indigenous Policy Journal, 3*(3). https://doi.org/10.18584/iipj.2012.3.3.10. Retrieved from https://ir.lib.uwo.ca/iipj/vol3/iss3/10.

Lane, S. N. (2014). Acting, predicting and intervening in a socio-hydrological world. *Hydrology and Earth System Sciences, 18*(3), 927–952.

Langton, M. (2002). The edge of the sacred, the edge of death: Sensual inscriptions. In B. David & M. Wilson (Eds.), *Inscribed Landscapes: Marking and Making Place* (pp. 253–269). Honolulu: University of Hawaii Press.

Larsen, S. C., & Johnson, J. T. (2017). *Being Together in Place: Indigenous Co-Existence in a More Than Human World.* Minneapolis, MN: University of Minnesota Press.

Lave, R. (2016). Stream restoration and the surprisingly social dynamics of science. *Wiley Interdisciplinary Reviews: Water, 3*(1), 75–81.

Lave, R. (2018). Stream mitigation banking. *Wiley Interdisciplinary Reviews: Water, 5*(3), e1279.

Lave, R., Robertson, M. M., & Doyle, M. W. (2008). Why you should pay attention to stream mitigation banking. *Ecological Restoration, 26*(4), 287–289.

Leopold, A. (1949). *A Sand County Almanac and Sketches Here and There.* Oxford: Oxford University Press.

Leopold, L. B. (1994). River morphology as an analog to Darwin's theory of natural selection. *Proceedings of the American Philosophical Society, 138*(1), 31–47.

Light, A. (2000). Ecological restoration and the culture of nature: A pragmatic perspective. In P. H. Gobster & R. B. Hull (Eds.), *Restoring Nature: Perspectives from the Social Sciences and Humanities* (pp. 49–70). Washington, DC: Island Press.

Linke, S., Gifford, T., Desjonquères, C., Tonolla, D., Aubin, T., Barclay, L., ... Sueur, J. (2018). Freshwater ecoacoustics as a tool for continuous ecosystem monitoring. *Frontiers in Ecology and the Environment, 16*(4), 231–238.

Linton, J. (2010). *What Is Water?: The History of a Modern Abstraction.* Vancouver, BC: UBC Press.

Linton, J., & Budds, J. (2014). The hydrosocial cycle: Defining and mobilizing a relational-dialectical approach to water. *Geoforum, 57,* 170–180.

Lockwood, A. (2007). What is a river. *Soundscape: The Journal of Acoustic Ecology, 7*(1), 43–44.

Longley, A., Fitzpatrick, K., Martin, R., Brown, C., Šunde, C., Ehlers, C., ... Waghorn, K. (2013). Imagining a fluid city. *Qualitative Inquiry, 19*(9), 736–740.

Lovelock, J. (2000). *Gaia: A New Look at Life on Earth.* Oxford: Oxford Paperbacks.

Lumsdon, A. E., Artamonov, I., Bruno, M. C., Righetti, M., Tockner, K., Tonolla, D., & Zarfl, C. (2018). Soundpeaking—Hydropeaking induced changes in river soundscapes. *River Research and Applications, 34*(1), 3–12.

Macklin, M. G., & Lewin, J. (2019). River stresses in anthropogenic times: Large-scale global patterns and extended environmental timelines. *Progress in Physical Geography: Earth and Environment, 43,* 3–23.

Maclean, K., Bark, R., Moggridge, B., Jackson, S., & Pollino, C. (2012). *Ngemba Water Values and Interests: Ngemba Old Mission Billabong and Brewarrina Aboriginal Fish Traps (Baiame's Nguunhu).* Canberra, ACT: CSIRO.

Mahood, K. (2016). *Position Doubtful. Mapping Landscapes and Memories.* Melbourne, VIC: Scribe.

Marr, A. (2012). *A History of the World.* London: Macmillan.

Marsh, G. P. (1864). *Man and Nature: Or Physical Geography as Modified by Human Action.* New York: Scribner.

McCully, P. (1996). *Silenced Rivers: The Ecology and Politics of Large Dams.* London: Zed Books.

McNiven, I., Crouch, J., Richards, T., Sniderman, K., Dolby, N., & Mirring, G. (2015). Phased redevelopment of an ancient Gunditjmara fish trap over the past 800 years: Muldoons Trap Complex, Lake Condah, Southwestern Victoria. *Australian Archaeology, 81*(1), 44–58.

McPhee, J. (1989). *The Control of Nature.* New York: Farrar, Straus & Giroux.

Meadows, D. H., Meadows, D. H., Randers, J., & Behrens, W. W., III. (1972). *The Limits to Growth: A Report for the Club of Rome's Project on the Predicament of Mankind*. New York: Universe Books.

Meppem, T., & Bourke, S. (1999). Different ways of knowing: A communicative turn toward sustainability. *Ecological Economics, 30*(3), 389–404.

Merchant, C. (1980). *The Death of Nature: Women. Ecology, and the Scientific Revolution*. San Francisco: Harper and Row.

Mika, S., Hoyle, J., Kyle, G., Howell, T., Wolfenden, B., Ryder, D., ... Fryirs, K. (2010). Inside the "black box" of river restoration: Using catchment history to identify disturbance and response mechanisms to set targets for process-based restoration. *Ecology and Society, 15*(4).

Millennium Ecosystem Assessment. (2005). *Ecosystems and Human Well-Being*. Washington, DC: Island Press.

Mitchell, D. (2003). Cultural landscapes: Just landscapes or landscapes of justice? *Progress in Human Geography, 27*(6), 787–796.

Molle, F. (2009). River-basin planning and management: The social life of a concept. *Geoforum, 40*(3), 484–494.

Monbiot, G. (2013). *Feral: Searching for Enchantment on the Frontiers of Rewilding*. London: Penguin.

Monbiot, G. (2019, April 15). Only rebellion will prevent an ecological collapse. *The Guardian*.

Montgomery, D. R. (2008). Dreams of natural streams. *Science, 319*(5861), 291–292.

Morin, K. M. (2009). Landscape: Representing and interpreting the world. In N. J. Clifford, S. L. Holloway, S. P. Rice, & G. Valentine (Eds.), *Key Concepts in Geography* (2nd ed., pp. 286–299). Los Angeles: Sage.

Morris, J. D., & Ruru, J. (2010). Giving voice to rivers: Legal personality as a vehicle for recognising Indigenous peoples' relationships to water? *Australian Indigenous Law Review, 14*(2), 49–62.

Moss, T., Medd, W., Guy, S., & Marvin, S. (2009). Organising water: The hidden role of intermediary work. *Water Alternatives, 2*(1), 16–33.

Mouat, C., Legacy, C., & March, A. (2013). The problem is the solution: Testing agonistic theory's potential to recast intractable planning disputes. *Urban Policy and Research, 31*(2), 150–166.

Mould, S. A., Fryirs, K., & Howitt, R. (2018). Practicing sociogeomorphology: Relationships and dialog in river research and management. *Society & Natural Resources, 31*(1), 106–120.

Muir, C. (2014). *The Broken Promise of Agricultural Progress*. London: Routledge. Environmental Humanities Series.

Muru-Lanning, M. (2016). *Tupuna Awa. People and Politics of the Waikato River* (p. 230). Auckland: University of Auckland Press.

Nash, R. (2001). *Wilderness and the American Mind*. New Haven, CT: Yale University Press.

Nassauer, J. I. (1995). Culture and changing landscape structure. *Landscape Ecology, 10*(4), 229–237.

Nestler, J. M., Pompeu, P. S., Goodwin, R. A., Smith, D. L., Silva, L. G., Baigun, C. R., & Oldani, N. O. (2012). The river machine: A template for fish movement and habitat, fluvial geomorphology, fluid dynamics and biogeochemical cycling. *River Research and Applications, 28*(4), 490–503.

Newson, M. D. (2010a). 'Catchment consciousness' – Will mantra, metric or mania best protect, restore and manage habitats? In *Atlantic Salmon Trust 40th Anniversary Conference*. Oxford: Blackwell.

Newson, M. (2010b). Understanding 'hot-spot' problems in catchments: The need for scale-sensitive measures and mechanisms to secure effective solutions for river management and conservation. *Aquatic Conservation: Marine and Freshwater Ecosystems, 20*(S1), 62–72.

Norgaard, R. B. (2006). *Development Betrayed: The End of Progress and a Co-Evolutionary Revisioning of the Future*. London: Routledge.

Norris, R. H., & Thoms, M. C. (1999). What is river health? *Freshwater Biology, 41*(2), 197–209.

Nucitelli, D. (2017, December 27). Fake news is a threat to humanity, but scientists may have a solution. *The Guardian*.

O'Connor, J. E., Duda, J. J., & Grant, G. E. (2015). 1000 dams down and counting. *Science, 348*(6234), 496–497.

O'Donnell, E. (2018). *Legal Rights for Rivers: Competition, Collaboration and Water Governance. Earthscan Studies in Water Resource Management*. New York: Routledge.

O'Donnell, E., & Talbot-Jones, J. (2018). Creating legal rights for rivers: Lessons from Australia, New Zealand, and India. *Ecology and Society, 23*(1).

O'Neil, C. (2016). *Weapons of Math Destruction: How Big Data Increases Inequality and Threatens Democracy*. London: Penguin.

O'Neill, J., Holland, A., & Light, A. (2008). *Environmental Values*. London: Routledge.

Olden, J. D., Konrad, C. P., Melis, T. S., Kennard, M. J., Freeman, M. C., Mims, M. C., … McMullen, L. E. (2014). Are large-scale flow experiments informing the science and management of freshwater ecosystems? *Frontiers in Ecology and the Environment, 12*(3), 176–185.

Oreskes, N., & Conway, E. M. (2011). *Merchants of Doubt: How a Handful of Scientists Obscured the Truth on Issues from Tobacco Smoke to Global Warming*. London: Bloomsbury Publishing.

Ostrom, E. (2009). A general framework for analyzing sustainability of social-ecological systems. *Science, 325*(5939), 419–422.

Oswald, A. (2012). *Dart*. London: Faber.

Palmer, T. (1994). *Lifelines. The Case for River Conservation.* Washington, DC: Island Press.

Palmer, M. A., Bernhardt, E. S., Allan, J. D., Lake, P. S., Alexander, G., Brooks, S., ... Galat, D. L. (2005). Standards for ecologically successful river restoration. *Journal of Applied Ecology, 42*(2), 208–217.

Palmer, M. A., Hondula, K. L., & Koch, B. J. (2014). Ecological restoration of streams and rivers: Shifting strategies and shifting goals. *Annual Review of Ecology, Evolution, and Systematics, 45,* 247–269.

Pannell, D. J., Roberts, A. M., Park, G., & Alexander, J. (2013). Improving environmental decisions: A transaction-costs story. *Ecological Economics, 88,* 244–252.

Park, G. (1995). *Ngā Uruora—The Groves of Life Ecology and History in a New Zealand Landscape.* Wellington: Victoria University Press.

Pascoe, B. (2014). *Dark Emu Black Seeds: Agriculture or Accident?* Broome, WA: Magabala Books.

Pauly, D. (1995). Anecdotes and the shifting baseline syndrome of fisheries. *Trends in Ecology & Evolution, 10*(10), 430.

Penck, A. (1897). Potamology as a branch of physical geography. *The Geographical Journal, 10*(6), 619–623.

Pessoa, F. (2002). *The Book of Disquiet.* London: Penguin.

Petts, G. E. (1984). *Impounded Rivers: Perspectives for Ecological Management.* Chichester: Wiley.

Petts, J. (2007). Learning about learning: Lessons from public engagement and deliberation on urban river restoration. *The Geographical Journal, 173*(4), 300–311.

Piégay, H., Chabot, A., & Le Lay, Y. F. (2019). Some comments about resilience: From cyclicity to trajectory, a shift in living and nonliving system theory. *Geomorphology.* https://doi.org/10.1016/j.geomorph.2018.09.018

Piégay, H., Darby, S. E., Mosselman, E., & Surian, N. (2005). A review of techniques available for delimiting the erodible river corridor: A sustainable approach to managing bank erosion. *River Research and Applications, 21*(7), 773–789.

Plumwood, V. (2002). *Feminism and the Mastery of Nature.* London: Routledge.

Poff, N. L. (2018). Beyond the natural flow regime? Broadening the hydro-ecological foundation to meet environmental flows challenges in a nonstationary world. *Freshwater Biology, 63*(8), 1011–1021.

Poff, N. L., Allan, J. D., Bain, M. B., Karr, J. R., Prestegaard, K. L., Richter, B. D., ... Stromberg, J. C. (1997). The natural flow regime. *BioScience, 47*(11), 769–784.

Postel, S., & Richter, B. (2003). *Rivers for Life: Managing Water for People and Nature.* Washington, DC: Island Press.

Prince, R. (2016). The spaces in between: Mobile policy and the topographies and topologies of the technocracy. *Environment and Planning D: Society and Space, 34*(3), 420–437.

Purdy, J. (2015). *After Nature: A Politics for the Anthropocene.* Cambridge, MA: Harvard University Press.

Pyle, M. T. (1995). Beyond fish ladders: Dam removal as a strategy for restoring America's rivers. *Stanford Environmental Law Journal, 14*(1), 97–143.

Rapport, D. J., Costanza, R., & McMichael, A. J. (1998). Assessing ecosystem health. *Trends in Ecology & Evolution, 13*(10), 397–402.

Reisner, M. (1986). *Cadillac Desert: The American West and Its Disappearing Water.* New York: Viking.

Relph, E. (1976). *Place and Placelessness.* London: Pion.

Robertson, M. M. (2006). The nature that capital can see: Science, state, and market in the commodification of ecosystem services. *Environment and Planning D: Society and Space, 24*(3), 367–387.

Rockström, J., Steffen, W. L., Noone, K., Persson, Å., Chapin, F. S., III, Lambin, E., … Nykvist, B. (2009). Planetary boundaries: Exploring the safe operating space for humanity. *Ecology and Society, 14*(2), 32.

Rogers, K. H. (2006). The real river management challenge: Integrating scientists, stakeholders and service agencies. *River Research and Applications, 22*(2), 269–280.

Rühs, N., & Jones, A. (2016). The implementation of earth jurisprudence through substantive constitutional rights of nature. *Sustainability, 8*(2), 174.

Ruru, J. (2018). Listening to Papatūānuku: A call to reform water law. *Journal of the Royal Society of New Zealand, 48,* 215–224.

Salmond, A. (2014). Tears of Rangi: Water, power, and people in New Zealand. *HAU: Journal of Ethnographic Theory, 4*(3), 285–309.

Salmond, A. (2017). *Tears of Rangi: Experiments Across Worlds.* Auckland: Auckland University Press.

Schama, S. (1995). *Landscape and Memory.* New York: Knopf.

Schmidt, J. C., Webb, R. H., Valdez, R. A., Marzolf, G. R., & Stevens, L. E. (1998). Science and values in river restoration in the Grand Canyon: There is no restoration or rehabilitation strategy that will improve the status of every riverine resource. *BioScience, 48*(9), 735–747.

Schulze-Makuch, D., & Bains, W. (2017). *The Cosmic Zoo: Complex Life on Many Worlds.* New York: Springer.

Shapin, S. (1998). Placing the view from nowhere: Historical and sociological problems in the location of science. *Transactions of the Institute of British Geographers, 23*(1), 5–12.

Shiva, V. (2002). *Water Wars: Privatization, Pollution, and Profit.* Cambridge, MA: South End Press.

Simon, A., Doyle, M., Kondolf, M., Shields, F. D., Jr., Rhoads, B., & McPhillips, M. (2007). Critical evaluation of how the Rosgen classification and associated "natural channel design" methods fail to integrate and quantify fluvial processes and channel response. *Journal of the American Water Resources Association (JAWRA), 43*(5), 1117–1131.

Singer, P. (1972). Famine, affluence, and morality. *Philosophy and Public Affairs, 1*(3), 229–243.

Slater, L. (2013). 'Wild rivers, wild ideas': Emerging political ecologies of Cape York Wild Rivers. *Environment and Planning D: Society and Space, 31*(5), 763–778.

Smith, J. L. (2017). I, River?: New materialism, riparian non-human agency and the scale of democratic reform. *Asia Pacific Viewpoint, 58*(1), 99–111.

Smith, B., Clifford, N. J., & Mant, J. (2014). The changing nature of river restoration: Changing nature of river restoration. *Wiley Interdisciplinary Reviews: Water, 1*(3), 249–261.

Smith, R. F., Hawley, R. J., Neale, M. W., Vietz, G. J., Diaz-Pascacio, E., Herrmann, J., … Utz, R. M. (2016). Urban stream renovation: Incorporating societal objectives to achieve ecological improvements. *Freshwater Science, 35*(1), 364–379.

Spink, A., Fryirs, K., & Brierley, G. (2009). The relationship between geomorphic river adjustment and management actions over the last 50 years in the upper Hunter catchment, NSW, Australia. *River Research and Applications, 25*(7), 904–928.

Spink, A., Hillman, M., Fryirs, K., Brierley, G., & Lloyd, K. (2010). Has river rehabilitation begun? Social perspectives from the Upper Hunter catchment, New South Wales, Australia. *Geoforum, 41*(3), 399–409.

Steffen, W., Broadgate, W., Deutsch, L., Gaffney, O., & Ludwig, C. (2015). The trajectory of the Anthropocene: The great acceleration. *The Anthropocene Review, 2*(1), 81–98.

Steffen, W., Grinevald, J., Crutzen, P., & McNeill, J. (2011). The Anthropocene: Conceptual and historical perspectives. *Philosophical Transactions of the Royal Society A: Mathematical, Physical and Engineering Sciences, 369*(1938), 842–867.

Stone, C. D. (1973). *Should Trees Have Standing? Toward Legal Rights for Natural Objects.* Los Altos, CA: William Kaufmann.

Strang, V. (2004). *The Meaning of Water.* Oxford: Berg.

Tadaki, M., Brierley, G., & Cullum, C. (2014). River classification: Theory, practice, politics. *Wiley Interdisciplinary Reviews: Water, 1*(4), 349–367.

Tadaki, M., & Sinner, J. (2014). Measure, model, optimise: Understanding reductionist concepts of value in freshwater governance. *Geoforum, 51*, 140–151.

Taylor, B., Van Wieren, G., & Zaleha, B. D. (2016). Lynn White Jr. and the greening-of-religion hypothesis. *Conservation Biology, 30*(5), 1000–1009.

Te Aho, L. (2010). Indigenous challenges to enhance freshwater governance and management in Aotearoa New Zealand—The Waikato river settlement. *The Journal of Water Law*, *20*(5), 285–292.

Thomas, A. C. (2015). Indigenous more-than-humanisms: Relational ethics with the Hurunui River in Aotearoa New Zealand. *Social & Cultural Geography*, *16*(8), 974–990.

Thomas, C. D. (2017). *Inheritors of the Earth*. London: Allen Lane, Penguin Random House.

Thornton, J., & Goodman, M. (2017). *Client Earth*. London: Scribe.

Tockner, K., Bernhardt, E. S., Koska, A., & Zarfl, C. (2016). A global view on future major water engineering projects. In R. F. Hüttl, O. Bens, C. Bismuth, & S. Hoechstetter (Eds.), *Society-Water-Technology* (pp. 47–64). Cham: Springer.

Tousley, N. (2018). Tanya Harnett: The poetics & politics of scarred/sacred water. In J. Ellis (Ed.), *Water Rites: Reimagining Water in the West* (pp. 33–43). Calgary: Calgary University Press.

Tsing, A. L. (2004). *Friction*. Princeton, NJ: Princeton University Press.

Tuan, Y. F. (1974). *Topophilia: A Study of Environmental Perceptions, Attitudes, and values*. New York: Columbia University Press.

Tuan, Y. F. (1977). *Space and Place: The Perspective of Experience*. Minneapolis, MN: University of Minnesota Press.

UN (WWAP (United Nations World Water Assessment Programme)/UN-Water). (2018). *The United Nations World Water Development Report 2018: Nature-Based Solutions for Water*. Paris: UNESCO.

Vörösmarty, C. J., McIntyre, P. B., Gessner, M. O., Dudgeon, D., Prusevich, A., Green, P., … Davies, P. M. (2010). Global threats to human water security and river biodiversity. *Nature*, *467*(7315), 555–561.

Vugteveen, P., Leuven, R. S. E. W., Huijbregts, M. A. J., & Lenders, H. J. R. (2006). Redefinition and elaboration of river ecosystem health: Perspective for river management. *Hydrobiologia*, *565*, 289–308.

Walker, W. E., Loucks, D. P., & Carr, G. (2015). Social responses to water management decisions. *Environmental Processes*, *2*(3), 485–509.

Walter, R. C., & Merritts, D. J. (2008). Natural streams and the legacy of water-powered mills. *Science*, *319*(5861), 299–304.

White, L. (1967). The historical roots of our ecologic crisis. *Science*, *155*(3767), 1203–1207.

White, R. (1990). Environmental history, ecology, and meaning. *The Journal of American History*, *76*(4), 1111–1116.

Wilcock, D., Brierley, G., & Howitt, R. (2013). Ethnogeomorphology. *Progress in Physical Geography*, *37*(5), 573–600.

Wilder, M., & Ingram, H. (2018). Knowing equity when we see it: Water equity in contemporary global contexts. In K. Conca & E. Weinthal (Eds.), *The Oxford Handbook of Water Politics and Policy*. Oxford: Oxford University Press.

Wilkinson, J. L., Hooda, P. S., Swinden, J., Barker, J., & Barton, S. (2018). Spatial (bio) accumulation of pharmaceuticals, illicit drugs, plasticisers, perfluorinated compounds and metabolites in river sediment, aquatic plants and benthic organisms. *Environmental Pollution, 234*, 864–875.

Wilson, E. O. (1984). *Biophilia*. Cambridge: Harvard University Press.

Wilson, G., & Lee, D. M. (2019). Rights of rivers enter the mainstream. *The Ecological Citizen, 2*(2), 183–187. Retrieved from https://www.ecologicalcitizen.net/pdfs/v02n2-13.pdf.

Wittfogel, K. A. (1955). Developmental aspects of hydraulic societies. In J. H. Steward (Ed.), *Irrigation Civilizations: A Comparative Study* (pp. 43–52). Washington, DC: Pan American Union.

Wohl, E., Angermeier, P. L., Bledsoe, B., Kondolf, G. M., MacDonnell, L., Merritt, D. M., … Tarboton, D. (2005). River restoration. *Water Resources Research, 41*(10).

Worster, D. (1993). *The Wealth of Nature: Environmental History and the Ecological Imagination*. Oxford: Oxford University Press.

Wright, R. (2004). *A Short History of Progress*. Toronto, ON: House of Anansi.

Wright, S. (2015). More-than-human, emergent belongings: A weak theory approach. *Progress in Human Geography, 39*(4), 391–411.

WWF (World Wildlife Fund). (2016). *Living Planet Report 2016. Risk and Resilience in a New Era*. Gland: WWF International.

Wylie, J. (2007). *Landscape*. Abingdon: Routledge.

Yates, J. S., Harris, L. M., & Wilson, N. J. (2017). Multiple ontologies of water: Politics, conflict and implications for governance. *Environment and Planning D: Society and Space, 35*(5), 797–815.

Yu, G. A., Huang, H. Q., Wang, Z., Brierley, G., & Zhang, K. (2012). Rehabilitation of a debris-flow prone mountain stream in southwestern China—Strategies, effects and implications. *Journal of Hydrology, 411*, 231–243.

INDEX

© The Author(s) 2020
G. J. Brierley, *Finding the Voice of the River*,
https://doi.org/10.1007/978-3-030-27068-1